Collins

Cambridge Lower Secondary

Maths

PROGRESS BOOK 9:
TEACHER PACK

Author: Alastair Duncombe

William Collins' dream of knowledge for all began with the publication of his first book in 1819.
A self-educated mill worker, he not only enriched millions of lives, but also founded a flourishing publishing house.
Today, staying true to this spirit, Collins books are packed with inspiration, innovation and practical expertise.
They place you at the centre of a world of possibility and give you exactly what you need to explore it.

Collins. Freedom to teach.

Published by Collins

An imprint of HarperCollinsPublishers
The News Building, 1 London Bridge Street, London, SE1 9GF, UK

HarperCollinsPublishers
Macken House, 39/40 Mayor Street Upper, Dublin 1, D01 C9W8, Ireland

Browse the complete Collins catalogue at
collins.co.uk

© HarperCollins*Publishers* Limited 2024

10 9 8 7 6 5 4 3 2 1

ISBN 978-0-00-866715-3

British Library Cataloguing-in-Publication Data
A catalogue record for this publication is available from the British Library.

The questions, accompanying marks and mark schemes included in this resource have been written by the author and are for
guidance only. They do not replicate examination papers and the questions in this resource will not appear in your exams. In
examinations the way marks are awarded may be different. Any references to assessment and/or assessment preparation are
the author's interpretation of the syllabus requirements.

This text has not been through the endorsement process for the Cambridge Pathway. Any references or materials related
to answers, grades, papers or examinations are based on the opinion of the author. The Cambridge International Education
syllabus or curriculum framework associated assessment guidance material and specimen papers should always be referred to
for definitive guidance.

Author: Alastair Duncombe
Publisher: Elaine Higgleton
Product manager: Catherine Martin
Product developer: Saaleh Patel
Copyeditor: Eric Pradel
Proofreader: Tim Jackson
Cover designer: Gordon MacGilp
Cover illustrator: Ann Paganuzzi
Typesetter: Ken Vail Graphic Design
Production controller: Sarah Hovell
Printed and bound by Ashford Colour Press Ltd

MIX
Paper | Supporting
responsible forestry
FSC™ C007454
www.fsc.org

This book contains FSC™ certified paper and other controlled
sources to ensure responsible forest management.

For more information visit: www.harpercollins.co.uk/green

Content

Introduction

This *Stage 9 Progress Teacher Pack* (and the *Stage 9 Progress Student Book*) can be used to support the *Collins Cambridge Stage 9 Lower Secondary Maths course* or to supplement your own resources.

The *Progress Teacher Pack* contains
- six Assessment Tasks – each corresponding to 4 or 5 chapters in the Collins Cambridge Stage 9 Maths course
- two End of Book Tests: Paper 1 is a non-calculator paper and Paper 2 is a calculator-allowed paper
- student Self-assessment sheets for each of the Assessment Tasks and End of Book Tests
- mark schemes for each of the Assessment Tasks and for the End of Book Tests.

How to use the Progress resources

This photocopiable Teacher Pack contains a range of Assessment Tasks and Tests that are designed to be valuable and flexible formative and summative assessment resources. They can be used to identify strengths and weaknesses and to pinpoint how future teaching should be adjusted to ensure all students make good progress.

The six Assessment Tasks can be used as class tests or can be set for students to complete at home. Each Task includes a list of the topics being tested, and begins with some multiple-choice questions to build confidence and to check understanding of some of the key ideas. As the students work through each Task, the questions become increasingly challenging.

Some of the questions in each Task are written to address the Cambridge *Thinking and Working Mathematically* characteristics:
- Specialising and generalising
- Conjecturing and convincing
- Characterising and classifying
- Critiquing and improving.

These questions may involve more thought or may involve students justifying their answer by providing a clear explanation or working.

Each Task is designed to be marked out of 30 (if 4 chapters are covered) or 40 (if 5 chapters are covered). Teachers may wish to set a time limit of say 40 or 50 minutes for the Task if they are using it as a class test.

The End of Book tests assess objectives taught across the whole year. The style of the End of Book tests is otherwise the same as the Assessment Tasks, with a mixture of question styles and question difficulties, as well as the inclusion of some Thinking and Working Mathematically questions. These tests could be used for summative purposes as end of year examinations or as practice for students ahead of their examinations.

The *Progress Teacher Pack* includes clear mark schemes for each Task and for the End of Book Tests. These mark schemes contain notes on what should be seen for full marks to be awarded. They also set out how part marks can be awarded in a question where the full correct answer is not reached. In some questions, the mark schemes allow for 'follow through' marks to be awarded – these allow for students to score marks in the second part of a question if they have correctly made use of their wrong answer to the first part.

The student Self-assessment sheets give students the opportunity to reflect on their understanding. Students record the mark for each question in the grids and then use these to find how well they have done with the questions relating to each chapter (or, for the End of Book Tests, each mathematics strand). This allows students to then reflect on which parts of the test went well and which areas they found harder. Students could pick out particular chapters as strengths or weaknesses. They could also comment on their success with Thinking and Working Mathematically questions or how they did on calculator or non-calculator questions.

The Self-assessment sheets also prompt students to set two targets. Target setting can play an important role in formative assessment if the targets are considered carefully and are revisited periodically. Students may be tempted to state a quite general target. However, a more achievable (and therefore more useful) target would be something more specific. For example,

Less helpful targets…

To become more confident at decimals ✗

To avoid making needless errors ✗

More helpful targets…

To become more confident at dividing a decimal by a whole number ✓

To try to avoid making needless errors by underlining key words in the question ✓

Teachers can use the results of the test and the students' Self-assessment sheets to help them in future lesson planning. For example, if many students struggled with work linked to simplifying expressions, a teacher may wish to bear this in mind when planning their teaching of a related topic, such as solving equations – teachers could, for instance, include starter activities recapping the earlier work.

Key features: Assessment Tasks

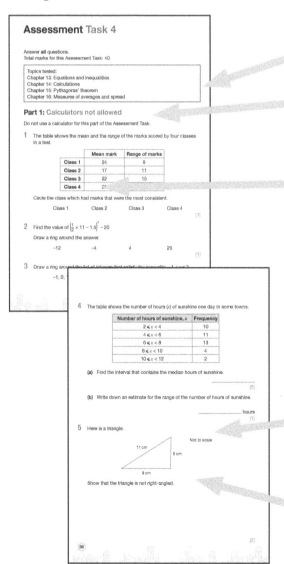

Clear indication of which topics are being tested in each Assessment Task.

A mixture of questions where calculators are allowed and not allowed.

Each Task begins with some multiple-choice questions to boost confidence and to assess key ideas.

Each Task contains some questions that relate to Thinking and Working Mathematically (TWM).

Clear layout and simple language. Space for working out.

Key features: Student self-assessment sheets

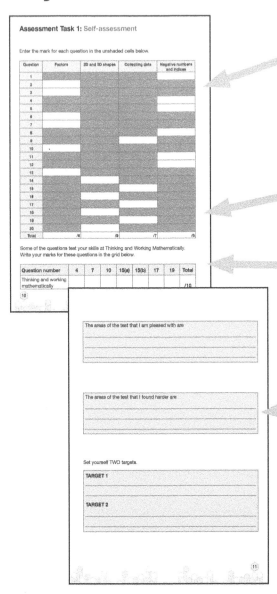

Clear table to link the marks scored on each question to the relevant textbook chapter/topic.

By totalling each column, students can compare how well they have scored in each topic area.

Success in the Thinking and Working Mathematically questions can also be analysed.

Space for students to reflect on how they have done and set targets.

Key features: Mark schemes

Clear indication of the correct answer and number of marks to be awarded.

Further information given where required about what is expected for full marks.

Guidance given about awarding follow through marks.

Guidance given about how to award part marks.

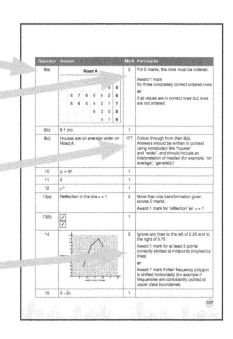

Assessment Task 1

Answer **all** questions.
Total marks for this Assessment Task: 30

> Topics tested:
> Chapter 1: Indices, roots and rational numbers
> Chapter 2: Angles
> Chapter 3: Collecting and organising data
> Chapter 4: Standard form

Part 1: Calculators not allowed

Do not use a calculator for this part of the Assessment Task.

1 Draw a ring around the value of 4200 correctly written in standard form.

42×10^2 42×10^3 4.2×10^3 4.2×10^2

[1]

2 Draw a ring around the best estimate of $\sqrt[3]{120}$

4   5   6   40

[1]

3 $7^{-4} \times 7^{-1} = 7^n$

Draw a ring around the value of n.

−5   −4   4   5

[1]

4 Match each calculation to its answer.

80×10^{-1}	80
$0.8 \div 10^{-2}$	8
$8 \div 10^3$	0.008

[1]

5 Write the following numbers in order of size, starting with the smallest.

8×10^5 $\qquad\qquad$ 9.2×10^4 $\qquad\qquad$ 7.5×10^5

_____ _____ _____

$\quad$ smallest $\qquad\qquad\qquad\qquad\qquad\qquad\qquad\qquad$ largest

[1]

6 Write each of the following without using indices.

The first one has been done for you.

$8^0 =$ _____1_____

$4^{-1} =$ _____

$2^{-4} =$ _____

[1]

7 Complete the statements by writing a whole number under each root sign.

$\sqrt{\rule{3em}{0pt}}$ $\quad$ is a rational number

$\sqrt{\rule{3em}{0pt}}$ $\quad$ is an irrational number

[1]

8 **(a)** Write 0.00467 in standard form.

_____

[1]

(b) Write 5.065×10^6 as an ordinary number.

_____

[1]

9 Mo collects toy vehicles.

Some of his vehicles are cars, some vehicles are red and some of his vehicles are old.

The Venn diagram shows the number of vehicles of each type.

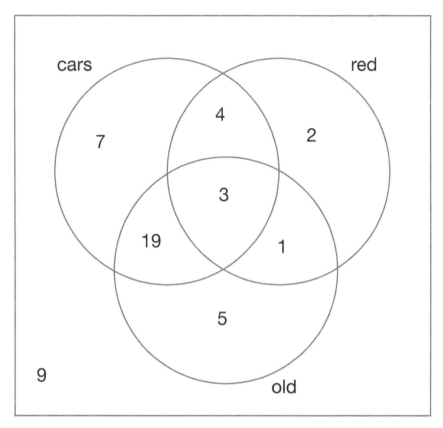

(a) Find the number of old cars in Mo's collection.

_____

[1]

(b) Complete the table to show the information.

	Cars		Not cars	
	Red	**Not red**	**Red**	**Not red**
Old			1	
Not old	4			

[2]

10 Laxmi estimates the value of $\sqrt{45}$ as 7.1

Comment on Laxmi's estimate.

_____

_____

[1]

11 m and n are numbers such that:

$m \times 10^{-2} = 7.5$

$5.5 \times 10^n = 0.0055$

Find the value of $m \div 10^n$

_____

[2]

Part 2: Calculators allowed

You may use a calculator for this part of the Assessment Task.

12 Draw a ring around the number that is written in standard form.

9.6×10^{-2} 12×10^{3} 65×10^{-4} 0.8×10^{5}

[1]

13 Draw a ring around the calculation that has an answer that is irrational.

$\frac{1}{3} \times 2$ $\frac{1}{3} - 1$ $\pi - \pi$ $\pi \times 2$

[1]

14 The diagram shows a quadrilateral with its exterior angles marked.

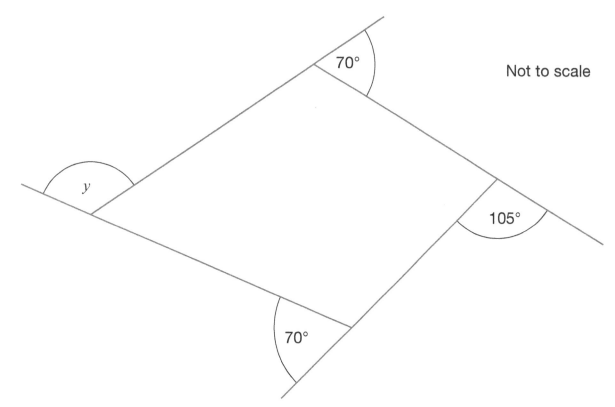

Draw a ring around the value of angle y.

115°   105°   75°   65°

[1]

15 A charity records the value of the donations that it receives one day.

$8.50	$4.25	$13.75	$12	$35
$3.10	$16.50	$10.45	$9.60	$17
$5.90	$2.80	$20	$14.90	$7.80

The charity's manager designs the following table for summarising these donations.

Donation value ($)	Frequency
0–19.99	
20–39.99	
40–59.99	
60–79.99	
80–99.99	

(a) Explain why the intervals in the table are not suitable for the data.

_____

_____

[1]

(b) Design a table with more suitable intervals.

Your table should have intervals of equal width.

You may not need to use every row in this table, and you don't need to complete the frequency column.

Donation value ($)	Frequency

[1]

16 A regular polygon has 16 sides.

Calculate the size of each interior angle.

_____ °

[2]

17 Here are three questions on a questionnaire.

Question 1 What is your age? _____

Question 2 How long did you spend exercising yesterday?

0–29 minutes ☐

30–59 minutes ☐

60 minutes or more ☐

Question 3 What do you think about the opening hours of the swimming pool?

_____

_____

Match each question number to the type of data.

Question 1

Question 2

Question 3

Qualitative data

Quantitative data

[1]

The diagram shows a hexagon with one of its exterior angles marked.

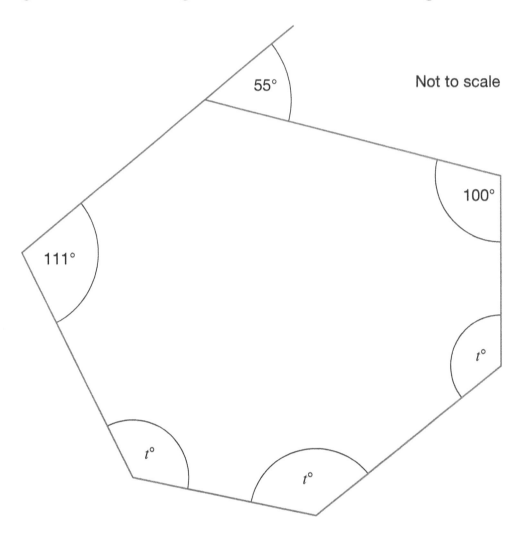

55°

Not to scale

100°

111°

$t°$

$t°$

$t°$

Calculate the value of t.

$t =$ _____

[3]

19 Robbie owns a shoe shop in a town.

He is investigating how often people in the town visit his shop.
He asks 50 of his customers.

Explain why Robbie's results are likely to be biased.

_____

_____

_____

[1]

20 The diagram shows an equilateral triangle *ABD*.

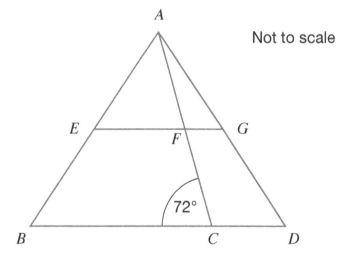

Not to scale

EFG and *BCD* are parallel straight lines.
Angle *FCB* = 72°

Calculate the size of angle *GAF*.

_____ °

[3]

Total marks: $\dfrac{}{30}$

Assessment Task 1: Self-assessment

Enter the mark for each question in the unshaded cells below.

Question	Indices, roots and rational numbers	Angles	Collecting and organising data	Standard form
1				
2				
3				
4				
5				
6				
7				
8				
9				
10				
11				
12				
13				
14				
15				
16				
17				
18				
19				
20				
Total	/6	/9	/7	/8

Some of the questions test your skills at Thinking and Working Mathematically. Write your marks for these questions in the grid below.

Question number	4	7	10	15(a)	15(b)	17	19	Total
Thinking and working mathematically								/7

The areas of the test that I am pleased with are

_____

_____

_____

The areas of the test that I found harder are

_____

_____

_____

Set yourself TWO targets.

TARGET 1

_____

_____

TARGET 2

_____

_____

Assessment Task 2

Answer **all** questions.
Total marks for this Assessment Task: 30
You will need mathematical instruments.
Tracing paper may be used.

Topics tested:
Chapter 5: Expressions
Chapter 6: Transformations
Chapter 7: Presenting and interpreting data 1
Chapter 8: Rounding and decimals

Part 1: Calculators not allowed

Do not use a calculator for this part of the Assessment Task.

1   Triangle Q is an enlargement of triangle P.

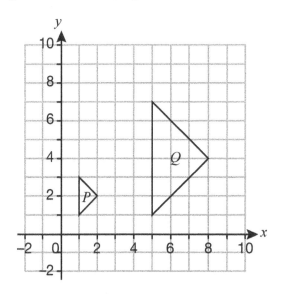

Draw a ring around the centre of enlargement.

(–1, 1)   (0, 1)   (–1, 0)   (0, 0)

[1]

2   An apple has mass 112 grams to the nearest gram.

Draw a ring around the lower bound for the mass of the apple.

111 grams   111.5 grams   111.9 grams   112 grams

[1]

3 Find the value of $\frac{(x-5)^2}{2}$ when $x = 1$

Draw a ring around the answer.

   −8   −4   4   8

[1]

4 Tick (✓) to show if the answer to each calculation will be greater than 12 or less than 12.

	Greater than 12	**Less than 12**
12 × 2.46	☐	☐
12 × 0.46	☐	☐
12 ÷ 0.46	☐	☐

[1]

5 Simplify $x^{12} \div x^4$

_____

[1]

6 Calculate 5.86 × 2.7

_____

[2]

7 The diagram shows a triangle T.

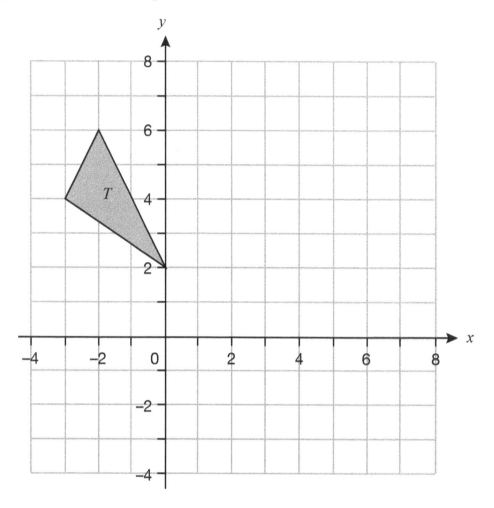

Triangle T is translated by vector $\begin{pmatrix} 5 \\ 1 \end{pmatrix}$ to give triangle U.

Triangle U is then rotated by 90° anticlockwise, centre (4, 1), to give triangle V.

Draw triangle V on the grid.

[2]

8 Johann expands the following brackets.

$(x + 6)(x + 7) = x^2 + 42$

$(y + 4)(y - 4) = y^2 - 16$

$(z + 5)(z - 7) = z^2 - 35$

Two of Johann's answers are wrong.

Tick (✓) or cross (✗) each answer in the table below.
Write correct simplified answers for the two questions he got wrong.

Johann's answer	✓ or ✗	Correct answer
$(x + 6)(x + 7) = x^2 + 42$		
$(y + 4)(y - 4) = y^2 - 16$		
$(z + 5)(z - 7) = z^2 - 35$		

[2]

9 Jana measures the width of the houses on two roads.

The back-to-back stem-and-leaf diagram shows the widths (in metres) for houses on Road B.

Road A					Road B			
				4	5	5	8	8
				5	2	2	9	
				6	0	2	4	8
				7	4	7	9	
				8	5			
				9	2			

Key: 4 | **5** | 2 represents a width of 5.4 metres in Road A
and a width of 5.2 metres in Road B.

(a) The widths (in metres) of the houses on Road A are:

6.7	8.0	5.9	9.4	7.6	8.5
7.1	6.6	6.4	7.8	8.3	7.4
9.1	6.5	7.3	6.2	7.5	6.8

Complete the stem-and-leaf diagram.

[2]

(b) Find the median width of a house on Road B.

_____ m

[1]

(c) The median width of a house on Road A is 7.35 m.

Use your answer to **(b)** to compare the width of the houses on Road A with those on Road B.

_____

_____

[1]

Part 2: Calculators allowed

You may use a calculator for this part of the Assessment Task.

10 Frankie thinks of a number y.

She first adds 4
She then squares the answer.

Draw a ring around the expression that represents the number she finishes with.

$y + 4^2$ $\qquad\qquad$ $y^2 + 4$ $\qquad\qquad$ $(y + 4)^2$ $\qquad\qquad$ $(y + 2)^2$

[1]

11 The back-to-back stem-and-leaf diagram shows the ages of people watching a film in two cinemas.

	Cinema 1				Cinema 2			
	9	8	8	**1**	7	9		
9	5	3	3	0	**2**	1	4	5	5
	5	4	2	**3**	2	6		
	8	1	0	**4**	4	5	8	
			2	**5**	3	4		
	8	8	5	**6**	4			

Key: 8 | **1** | 7 represents an age of 18 years in Cinema 1 and an age of 17 years in Cinema 2

Draw a ring around the number of people in Cinema 1 who are older than 45 years.

4 $\qquad\qquad$ 5 $\qquad\qquad$ 6 $\qquad\qquad$ 9

[1]

12 Simplify $(y^2)^5$

Draw a ring around the answer.

y^7 $\qquad\qquad$ y^{10} $\qquad\qquad$ y^{25} $\qquad\qquad$ y^{32}

[1]

13 (a) Quadrilaterals E and F are shown on the grid.

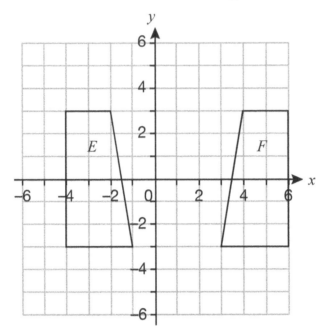

Describe the single transformation that maps quadrilateral E to quadrilateral F.

_____

_____

[2]

(b) A triangle is acted on by the combinations of transformations described on Card A and on Card B.

Card A
Translation by vector $\begin{pmatrix} 3 \\ 4 \end{pmatrix}$ and then reflection in the line $x = 2$

Card B
Rotation by 180° centre (4, 0) and then rotation by 90° anticlockwise centre (0, 0)

Tick (✓) the correct statement(s).

When the triangle is transformed by the transformations on Card A, the angles in the image will be the same as the angles in the original triangle. ☐

When the triangle is transformed by the transformations on Card B, the image will be the same size as the original triangle. ☐

[1]

14 The table shows the length (in minutes) of 30 songs.

Length of song (x minutes)	Frequency
$2 \leqslant x < 2.5$	3
$2.5 \leqslant x < 3$	9
$3 \leqslant x < 3.5$	11
$3.5 \leqslant x < 4$	7

Draw a frequency polygon to show the information.

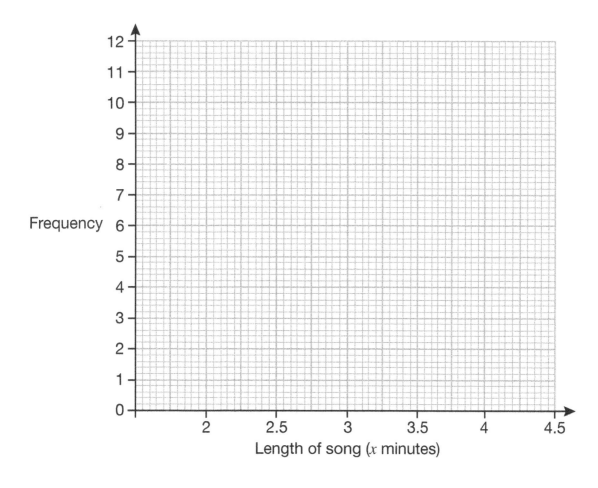

Length of song (x minutes)

[2]

15 Simplify $\dfrac{20 - 8x}{4}$

[1]

16 Pentagon P is shown on the grid.

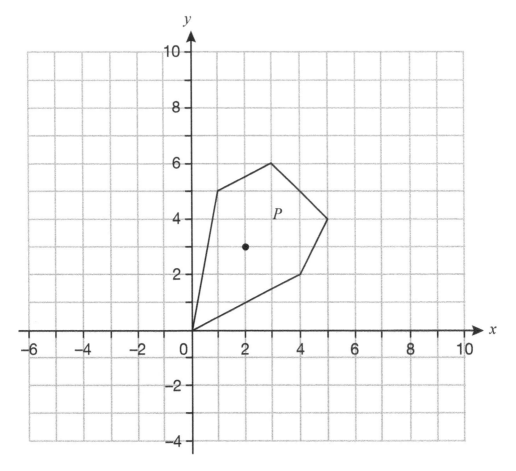

Enlarge pentagon P by scale factor 2, centre (2, 3).

[2]

17 A rectangle has length 12.4 cm and width 8.0 cm, each correct to the nearest millimetre.

Write down:

(a) the upper bound of the length

_____ cm

[1]

(b) the lower bound of the width.

_____ cm

[1]

18 Shapes L and M are regular polygons.

Shape M is an enlargement of shape L.

Shape L
Side length = 2.5 cm
Area = 48 cm²

Shape M
Side length = 12.5 cm
Area = ?

Show that the area of shape M is greater than 1000 cm².

[2]

Total marks: ___ / 30

Assessment Task 2: Self-assessment

Enter the marks for each question in the unshaded cells below.

Question	Expressions	Transformations	Presenting and interpreting data 1	Rounding and decimals
1				
2				
3				
4				
5				
6				
7				
8				
9				
10				
11				
12				
13				
14				
15				
16				
17				
18				
Total	/7	/10	/7	/6

Some of the questions test your skills at Thinking and Working Mathematically. Write your marks for these questions in the grid below.

Question number	4	8	13(b)	18	Total
Thinking and working mathematically					/6

The areas of the test that I am pleased with are

_____

_____

_____

The areas of the test that I found harder are

_____

_____

_____

Set yourself TWO targets.

TARGET 1

_____

_____

TARGET 2

_____

_____

Assessment Task 3

Answer **all** questions.
Total marks for this Assessment Task: 30
You will need mathematical instruments.

> Topics tested:
> Chapter 9: Functions and formulae
> Chapter 10: Accurate drawing
> Chapter 11: Fractions
> Chapter 12: Probability 1

Part 1: Calculators not allowed

Do not use a calculator for this part of the Assessment Task.

1 Work out $1\frac{1}{2} \times \frac{1}{2}$

Draw a ring around the answer.

$1\frac{1}{4}$ $\qquad$ $\frac{3}{4}$ $\qquad$ $\frac{3}{2}$ $\qquad$ $\frac{1}{4}$

[1]

2 Draw a ring around the fraction that is equivalent to a recurring decimal.

$\frac{7}{8}$ $\qquad$ $\frac{17}{20}$ $\qquad$ $\frac{19}{25}$ $\qquad$ $\frac{23}{30}$

[1]

3 Here is a function.

$$y = x^2 + 6$$

Draw a ring around the output y when the input is $x = 3$

12 $\qquad$ 15 $\qquad$ 18 $\qquad$ 81

[1]

4 A six-sided dice is rolled 180 times.

(a) Work out the expected number of times the dice would land on 5 if the dice is fair.

_____

[1]

(b) The dice actually lands on 5 in 10 of the 180 rolls.

Tick (✓) to show if you think the dice is fair or not.

The dice is likely to be fair. ☐   The dice is unlikely to be fair. ☐

Give a reason for your answer.

_____

_____

[1]

5 Calculate $\frac{3}{4} \div 1\frac{1}{5}$

Give your answer as a fraction in its simplest form.

_____

[2]

6 The scale drawing shows the position of two villages, C and D.

The scale is 1 cm represents 1 km.

Another village, E, is 9.4 km from D, on a bearing of 096° from D.

(a) Mark the position of E on the scale drawing.

[2]

(b) Measure the bearing of E from C.

_____

[1]

7 Rearrange the formula $h = \dfrac{5g - 7}{6}$ to make g the subject.

_____

[2]

8 Show that $\left(3\frac{2}{3} - 1\frac{8}{9}\right) \times 2\frac{1}{4}$ simplifies to a whole number.

_____

[3]

Part 2: Calculators allowed

You may use a calculator for this part of the Assessment Task.

9 A fruit bowl contains 15 pieces of fruit.
There are 2 bananas, 3 apples and 4 oranges.
The remaining pieces of fruit are mangos.
A piece of fruit is taken from the bowl at random.

Draw a ring around the probability that the piece of fruit is a mango.

$\frac{1}{4}$ $\qquad\qquad$ $\frac{2}{3}$ $\qquad\qquad$ $\frac{1}{6}$ $\qquad\qquad$ $\frac{2}{5}$

[1]

10 A is the point (1, 5)
B is the point (7, 11)

Draw a ring around the coordinates of the point that is $\frac{1}{3}$ of the way from A to B.

(3, 7) $\qquad\qquad$ (4, 8) $\qquad\qquad$ (6, 6) $\qquad\qquad$ (2, 2)

[1]

11 Rearrange the formula $y = 2 + \sqrt{x}$ to make x the subject.

Draw a ring around the answer.

$x = y^2 - 2$ $\qquad$ $x = \sqrt{y-2}$ $\qquad$ $x = \sqrt{y} - 2$ $\qquad$ $x = (y-2)^2$

[1]

12 Do not rub out your construction arcs in this question.

(a) Construct an angle of 60° at A.

The construction has been started for you.

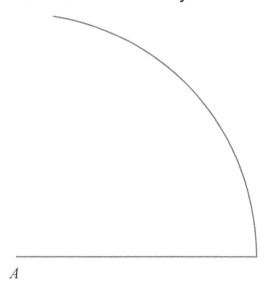

A

[1]

(b) Construct an equilateral triangle inscribed in the circle shown below.

The construction has been started for you.

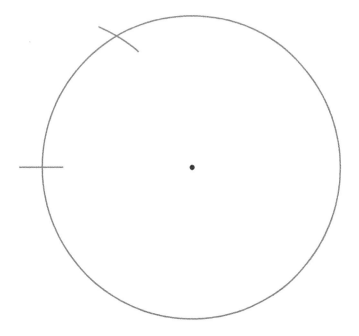

[2]

13 Here are six numbers.

$$3 \qquad 4 \qquad 8 \qquad 9 \qquad 12 \qquad 15$$

Use each of these numbers exactly **once** to make:

- two fractions that are equivalent to recurring decimals and
- one fraction equivalent to a terminating decimal.

Each fraction should be less than 1.

Equivalent to recurring decimals: $\dfrac{\square}{\square}$ and $\dfrac{\square}{\square}$

Equivalent to terminating decimal: $\dfrac{\square}{\square}$

[2]

14 (a) A bag contains a large number of balls.
Each ball is numbered with a single number: 1, 2, 3 or 4
A ball is picked at random from the bag.
The probability of getting an odd number is 0.6

Complete the table to show possible probabilities of picking each number.

Number	1	2	3	4
Probability				

[1]

(b) A spinner has three sections.

One is coloured red, a second is green and the third section is blue.
The table shows the probability that the spinner lands on red.

Colour	Red	Green	Blue
Probability	0.28		

The probability of the spinner landing on green is equal to the probability of it landing on blue.

Complete the table.

[2]

15 A function has the formula $y = 10x^2$

Find the value of the input x if:

  x is negative
  and
  $y = 90$

[2]

16 *ABC* is a right-angled triangle.

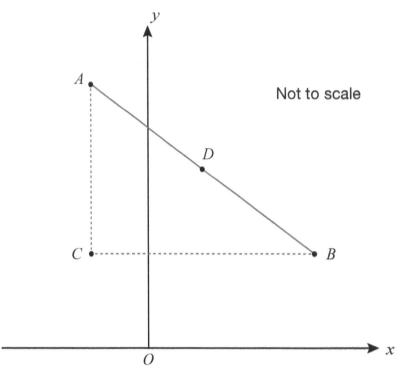

Not to scale

D is the midpoint of side *AB*.
A has coordinates (–4, 13) and *D* has coordinates (4, 9).

Find the coordinates of the point $\frac{3}{4}$ of the way along *CB*, starting from *C*.

(_____ , _____)

[2]

Total marks: $\dfrac{}{30}$

Assessment Task 3: Self-assessment

Enter the marks for each question in the unshaded cells below.

Question	Functions and formulae	Accurate drawing	Fractions	Probability 1
1				
2				
3				
4				
5				
6				
7				
8				
9				
10				
11				
12				
13				
14				
15				
16				
Total	/6	/9	/9	/6

Some of the questions test your skills at Thinking and Working Mathematically. Write your marks for these questions in the grid below.

Question number	4(b)	8	13	14(a)	Total
Thinking and working mathematically					/7

The areas of the test that I am pleased with are

_____

_____

_____

The areas of the test that I found harder are

_____

_____

_____

Set yourself TWO targets.

TARGET 1

_____

_____

TARGET 2

_____

_____

Assessment Task 4

Answer **all** questions.
Total marks for this Assessment Task: 30

Topics tested:
Chapter 13: Equations and inequalities
Chapter 14: Calculations
Chapter 15: Pythagoras' theorem
Chapter 16: Measures of averages and spread

Part 1: Calculators not allowed

Do not use a calculator for this part of the Assessment Task.

1 The table shows the mean and the range of the marks scored by four classes in a test.

	Mean mark	Range of marks
Class 1	24	9
Class 2	17	11
Class 3	22	15
Class 4	20	7

Circle the class which had marks that were the most consistent.

Class 1   Class 2   Class 3   Class 4

[1]

2 Find the value of $\left(\frac{1}{2} \times 11 - 1.5\right)^2 - 20$

Draw a ring around the answer.

−12   −4   4   29

[1]

3 Draw a ring around the list of integers that satisfy the inequality $-1 < x \leqslant 2$

−1, 0, 1   0, 1   0, 1, 2   −1, 0, 1, 2

[1]

4 The table shows the number of hours of sunshine x one day in some towns.

Number of hours of sunshine, x	Frequency
$2 \leqslant x < 4$	10
$4 \leqslant x < 6$	11
$6 \leqslant x < 8$	13
$8 \leqslant x < 10$	4
$10 \leqslant x < 12$	2

(a) Find the interval that contains the median hours of sunshine.

_____

[2]

(b) Write down an estimate for the range of the number of hours of sunshine.

_____ hours

[1]

5 Here is a triangle.

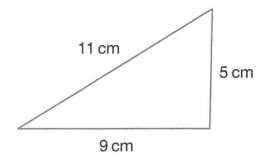

Not to scale

11 cm

5 cm

9 cm

Show that the triangle is not right-angled.

[2]

6 Show the inequality $-2 \leqslant x \leqslant 6$ on the number line.

[1]

7 **(a)** Sumi works out $\frac{1}{5} + 1.8 \times 10$

Here is her working.

$\frac{1}{5} + 1.8 = 0.2 + 1.8 = 2$

$2 \times 10 = 20$

So, $\frac{1}{5} + 1.8 \times 10 = 20$

Her answer is wrong.

Show how Sumi should have worked out $\frac{1}{5} + 1.8 \times 10$

[1]

(b) Calculate $4^2 \times 0.36 \times \frac{1}{8}$

_____

[2]

8 Solve $\dfrac{27}{2m-4} = 3$

$m =$ _____

[3]

Part 2: Calculators allowed

You may use a calculator for this part of the Assessment Task.

9 The frequency diagram represents the handspans (x centimetres) of a group of people.

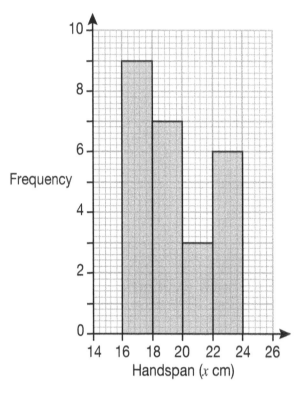

Draw a ring around the modal interval.

$16 \leqslant x < 18$ $\qquad$ $18 \leqslant x < 20$ $\qquad$ $20 \leqslant x < 22$ $\qquad$ $22 \leqslant x < 24$

[1]

10 Here is a right-angled triangle.

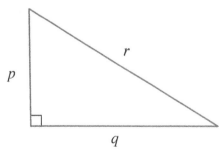

Draw a ring around the correct relationship between the side lengths.

$r = p + q$ $\qquad$ $r^2 = p^2 + q^2$ $\qquad$ $p = q + r$ $\qquad$ $p^2 = q^2 + r^2$

[1]

11 Solve $\frac{12}{n} = 3$

Draw a ring around the solution.

$n = 4$ $\qquad\qquad$ $n = \frac{1}{4}$ $\qquad\qquad$ $n = 36$ $\qquad\qquad$ $n = \frac{1}{36}$

[1]

12 Draw lines to match each inequality to the largest integer value that satisfies it.

$0 < x < 9$

$2x < 15$

$-2 < x + 3 \leqslant 10$

$12 - x > 2$

7

8

9

[1]

13 Here is a right-angled triangle.

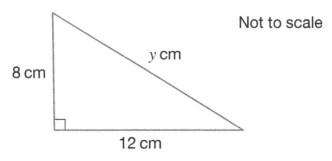

Not to scale

y cm

8 cm

12 cm

Find the value of y.

_____

[2]

14 The table shows the height (h metres) of a sample of trees growing in Park A.

Height (h metres)	Frequency
$5 \leqslant h < 10$	10
$10 \leqslant h < 15$	17
$15 \leqslant h < 20$	8
$20 \leqslant h < 25$	5

(a) Calculate an estimate of the mean height of the trees in the sample.

_____ m

[3]

(b) An estimate of the mean height of a sample of trees in Park B is 16.2 metres.

Lucille compares the heights of the trees in the two parks.

She says, "The mean for Park A is less than the mean for Park B."

Lucille's teacher suggests that she should write her comparison in context.

Write an improved comparison of the heights of the trees in the two parks.

_____

_____

_____

[1]

15 Solve the inequality $11 - 2x > 23 + 4x$

_____

[2]

16 _ABC_ and _ABD_ are right-angled triangles.

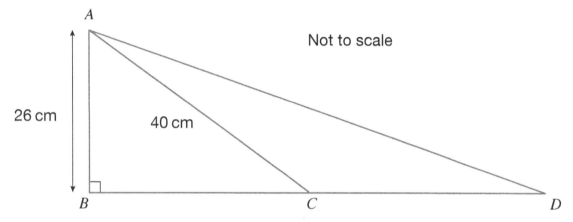

Not to scale

$AB = 26$ cm and $AC = 40$ cm.
BD is twice the length of BC.

Calculate the length of side AD.

_____ cm

[3]

Total marks: $\dfrac{\quad}{30}$

Assessment Task 4: Self-assessment

Enter the marks for each question in the unshaded cells below.

Question	Equations and inequalities	Calculations	Pythagoras' theorem	Measures of averages and spread
1				
2				
3				
4				
5				
6				
7				
8				
9				
10				
11				
12				
13				
14				
15				
16				
Total	/9	/4	/8	/9

Some of the questions test your skills at Thinking and Working Mathematically. Write your marks for these questions in the grid below.

Question number	5	7(a)	12	14(b)	Total
Thinking and working mathematically					/5

The areas of the test that I am pleased with are

_____

_____

_____

The areas of the test that I found harder are

_____

_____

_____

Set yourself TWO targets.

TARGET 1

_____

_____

TARGET 2

_____

_____

Assessment Task 5

Answer **all** questions.
Total marks for this Assessment Task: 40
You will need mathematical instruments.

> Topics tested:
> Chapter 17: Percentages
> Chapter 18: Sequences
> Chapter 19: Area and measures
> Chapter 20: Presenting and interpreting data 2
> Chapter 21: Ratio and proportion

Part 1: Calculators not allowed

Do not use a calculator for this part of the Assessment Task.

1   An old camera has a value of $500
    Its value is predicted to increase by 4% each year.

    Draw a ring around the calculation that gives the predicted value of the camera in 3 years' time.

    500×1.04^3 $500 \times 1.04 \times 3$ 500×0.04^3 $500 \times 0.04 \times 3$

    [1]

2   A sequence has first term 2
    The term-to-term rule is 'square and then add 1'

    Draw a ring around the 3rd term in the sequence.

    5   9   26   100

    [1]

3   A circle has a diameter of 8 cm.

    Draw a ring around the calculation that gives the area of the circle.

    $\pi \times 8$ $\pi \times 4$ $\pi \times 8^2$ $\pi \times 4^2$

    [1]

4 The dual bar chart shows information about the number of cars and vans parking in a car park on two days.

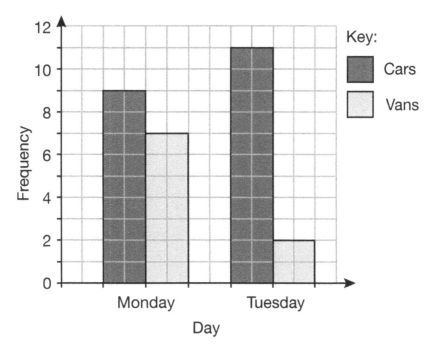

Draw a compound bar chart to show the same information.

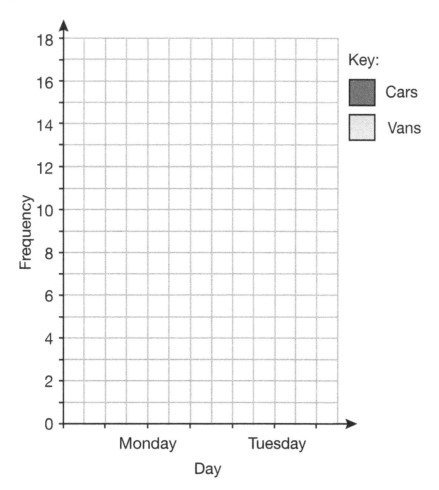

[3]

5 It takes 28 seconds to print a document 8 pages long.

Find how long it takes to print a document 12 pages long.

_____ seconds

[1]

6 Here are some statements about metric measurements.

A
1 kilogram > 1 gram

B
1 microgram > 1 gram

C
1 tonne > 1 kilogram

D
1 milligram > 1 microgram

E
1 milligram > 1 tonne

Write the letter for each statement in the correct column in the table.
The first one has been done for you.

Statement is correct	Statement is not correct
A	

[2]

7 Ten students each take a sports quiz and a history quiz.

The scatter diagram shows the marks they scored on each quiz.

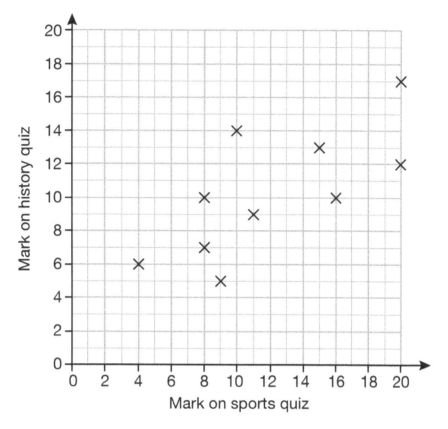

(a) Describe the relationship between the marks scored on the two quizzes.

_____

_____

[1]

(b) Draw a line of best fit on the scatter diagram.

[1]

(c) Another student scores 12 marks on the sports quiz.

Use your line of best fit to estimate the student's mark on the history quiz.

_____

[1]

8 A computer costs $800
Its price first increases by 20%
Then the new price is reduced by 10%

Find the cost of the computer after both percentage changes.

$ _____

[3]

9 Water is poured into two glasses.
The ratio of the amount of water in the two glasses is 3 : 5

One glass contains 300 ml of water.

Find the two possible values for the amount of water in the other glass.

_____ ml or _____ ml

[2]

10 Find an expression for the nth term for each sequence.
The first one has been done for you.

Sequence	nth term
$\frac{1}{7}, \frac{2}{7}, \frac{3}{7}, \frac{4}{7}, \ldots$	$\frac{n}{7}$
1, 8, 27, 64, . . .	
−4, −1, 4, 11, 20, . . .	

[3]

Part 2: Calculators allowed

You may use a calculator for this part of the Assessment Task.

11 Jan and Kim share an amount of money in the ratio 3 : 4
Jan receives $420

Draw a ring around the total amount of money that was shared.

$180 $560 $735 $980

[1]

12 The nth term of a sequence is given by the formula $n^3 - 1$

Draw a ring around the 2nd term of the sequence.

1   5   7   15

[1]

13 Here is a scatter graph.

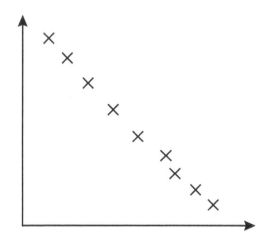

Draw a ring around the type of correlation shown on the scatter graph.

Weak positive   Strong positive

Weak negative   Strong negative

[1]

14 The diagram shows a circle with a radius of 5.5 cm.

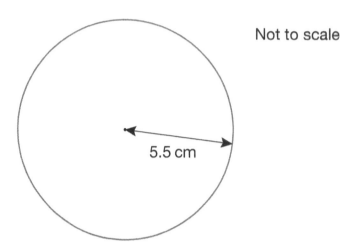

Not to scale

5.5 cm

Find the area of the circle.

_____ cm²

[1]

15 (a) Complete the conversion.

7 micrometres = _____ nanometres

[1]

(b) Write a unit on the answer line to make a correct inequality.

1 kilobyte < 1 _____ < 1 gigabyte

[1]

16 It takes 10 workers 72 hours to pick some apples.

Find how long it would take 6 workers, working at the same rate, to pick the same number of apples.

_____ hours

[1]

17 Students at a school study exactly one of the following subjects:

Music, Drama or Art.

The subjects studied by a group of 40 students at the school are shown in the bar chart.

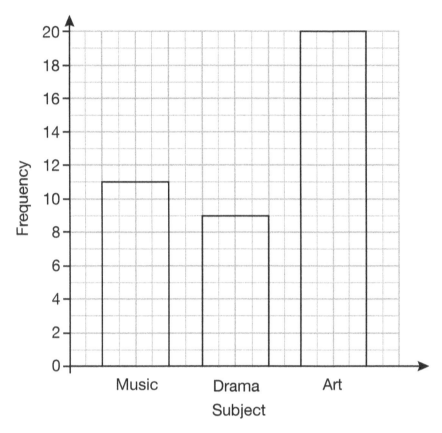

Draw a pie chart to show this information.

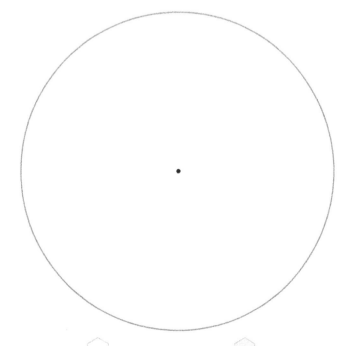

[3]

18 Find the perimeter of this semicircle.

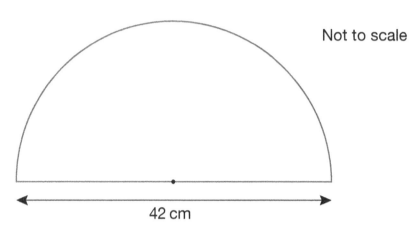

Not to scale

42 cm

_____ cm

[2]

19 (a) A sequence begins: 5, 7, 10, 14, 19
Find the next two terms in the sequence.

_____ and _____

[1]

(b) All the terms of a different sequence are positive integers.
The term-to-term rule is 'square and then subtract 18'
The 3rd term of this sequence is less than 100

Find the 1st term of this sequence.

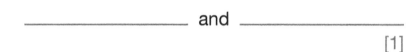

_____

[2]

20 The diagram shows a compound shape made from a right-angled triangle and a square.

A semicircle has been removed from the triangle.

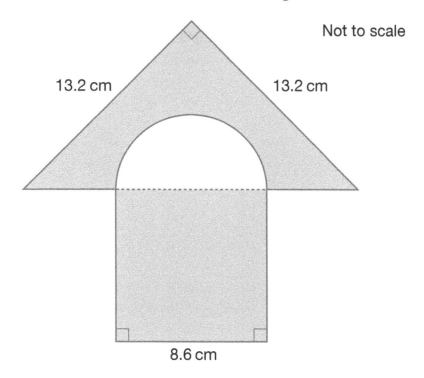

Not to scale

13.2 cm   13.2 cm

8.6 cm

Show that the shaded area is approximately 132 cm²

[3]

21   Huda buys an electric bike for $950

The value of Huda's bike decreases by 20% after 1 year.
After the first year, the value of her bike decreases by 12% every year.

Find the value of Huda's bike after 4 years.

$_____

[2]

Total marks: $\frac{\quad}{40}$

Assessment Task 5: Self-assessment

Enter the marks for each question in the unshaded cells below.

Question	Percentages	Sequences	Area and measures	Presenting and interpreting data 2	Ratio and proportion
1					
2					
3					
4					
5					
6					
7					
8					
9					
10					
11					
12					
13					
14					
15					
16					
17					
18					
19					
20					
21					
Total	/6	/8	/11	/10	/5

Some of the questions test your skills at Thinking and Working Mathematically. Write your marks for these questions in the grid below.

Question number	6	9	10	19(b)	20	Total
Thinking and working mathematically						/12

The areas of the test that I am pleased with are

_____

_____

_____

The areas of the test that I found harder are

_____

_____

_____

Set yourself TWO targets.

TARGET 1

_____

_____

TARGET 2

_____

_____

Assessment Task 6

Answer **all** questions.
Total marks for this Assessment Task: 40

> Topics tested:
> Chapter 22: Relationships and graphs
> Chapter 23: Probability 2
> Chapter 24: 3D shapes
> Chapter 25: Simultaneous equations
> Chapter 26: Thinking statistically

Part 1: Calculators not allowed

Do not use a calculator for this part of the Assessment Task.

1   Here is a graph.

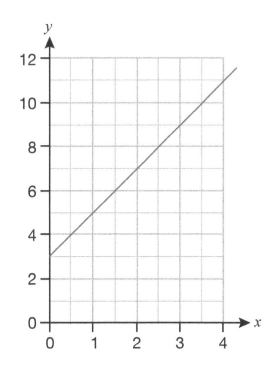

Draw a ring around the gradient of the graph.

$\frac{1}{2}$ 1   2   3

[1]

2 The diagram shows a cuboid.

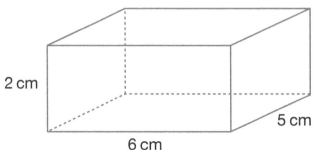

Not to scale

2 cm

5 cm

6 cm

Draw a ring around the number of planes of symmetry of this cuboid.

2   3   4   12

[1]

3 A dice is rolled and a coin is thrown.

Draw a ring around the probability of getting a 6 on the dice and tails on the coin.

$\frac{1}{6}$ $\frac{2}{3}$ $\frac{1}{12}$ $\frac{1}{36}$

[1]

4 The diagram shows a prism with a length of 10 cm.

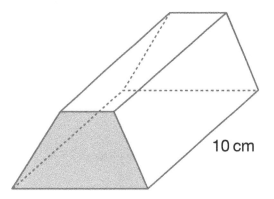

Not to scale

10 cm

The cross-sectional area of the prism is 18 cm^2

Calculate the volume of the prism.

_____ cm^3

[1]

5 A train company draws this graph to show how customer complaints have fallen.

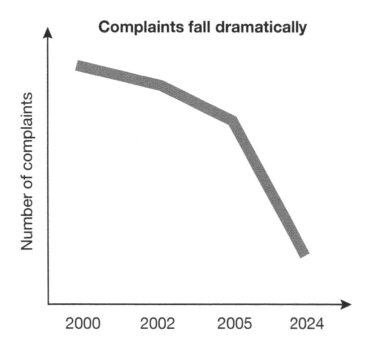

Write two reasons why this graph is misleading.

Reason 1 _____

_____

Reason 2 _____

_____

[2]

6 Solve.

$$2x + y = 28$$
$$x - y = 11$$

x = _____   y = _____

[2]

7 The probability that Fi has pizza for dinner is $\frac{1}{5}$

The probability that Fi reads after dinner is $\frac{4}{7}$

The tree diagram shows some of the probabilities.

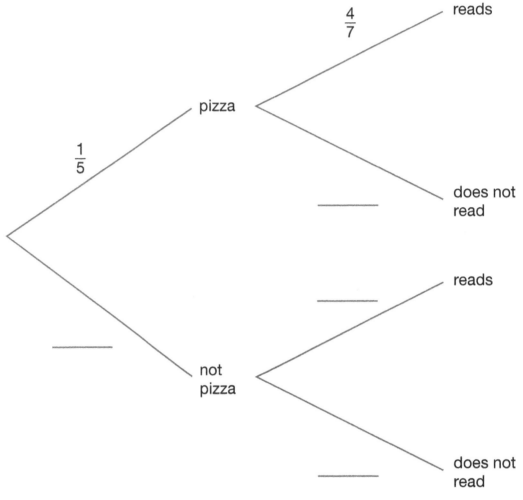

(a) Complete the tree diagram.
You may assume that Fi reading is independent of her having pizza.

[2]

(b) Find the probability that Fi has pizza for dinner and reads.

_____

[2]

8 **(a)** Complete this table of values for $y = x^2 - 6$

x	–2	–1	0	1	2	3	4
y		–5	–6			3	

[2]

(b) Draw the graph of $y = x^2 - 6$ for values of x between –2 and 4

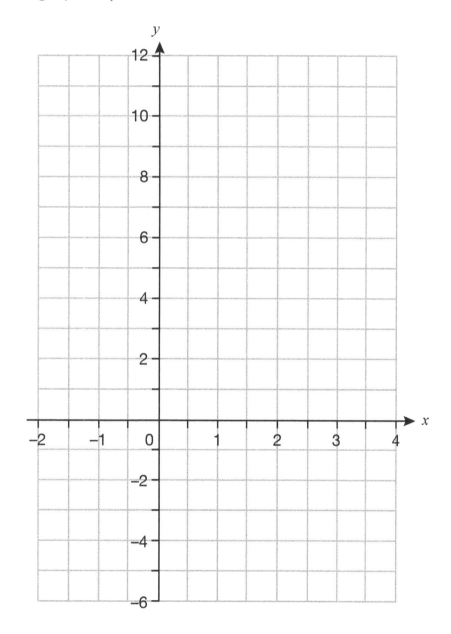

[2]

9 Kieron has a bag containing 2 green balls and 3 red balls.

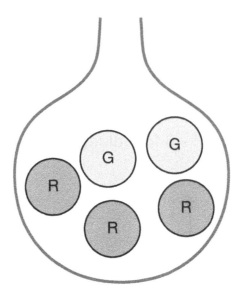

He takes a ball from the bag at random and does not replace it.
He then takes a second ball from the bag.

Kieron says, '*The probability that the second ball is red is independent of the colour of the first ball.*'

Tick (✓) to show if Kieron is correct or not.

He is correct. ☐   He is not correct. ☐

Give a reason for your answer.

_____

_____

[1]

10 The diagram shows a rectangle.

All measurements are in centimetres.

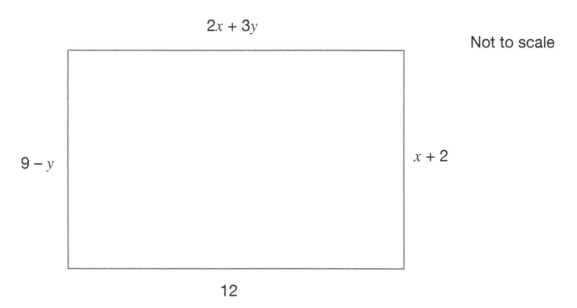

Not to scale

$2x + 3y$

$9 - y$

$x + 2$

12

By forming and solving simultaneous equations, find the area of the rectangle.

_____ cm²

[3]

Part 2: Calculators allowed

You may use a calculator for this part of the Assessment Task.

11 Three lines are shown on the grid.

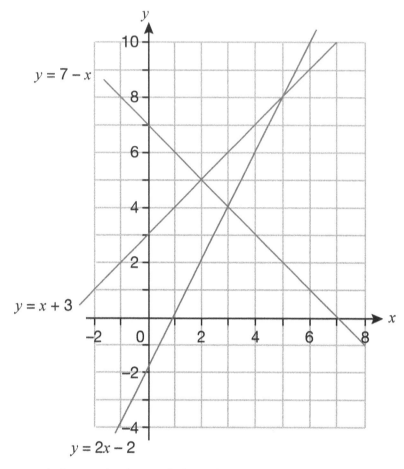

Draw a ring around the solution of the simultaneous equations:

$$y = 2x - 2$$
$$y = 7 - x$$

$x = 3, y = 4$ $x = 2, y = 5$ $x = 5, y = 2$ $x = 4, y = 3$

[1]

12 Apples are sold in small packs and large packs.
A small pack contains 4 apples and a large pack contains 10 apples.
Amol buys exactly 72 apples.
He buys x small packs and y large packs.

Draw a ring around the equation that describes this situation.

$\frac{x}{4} + \frac{y}{10} = 72$ $4x + 10y = 72$ $x + y = 72$ $x + 4 + y + 10 = 72$

[1]

13 The diagram shows a solid pyramid with a square base.

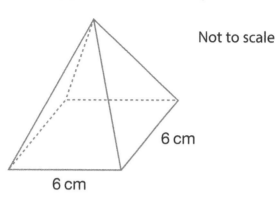

Not to scale

6 cm

6 cm

Each triangular face has an area of 25 cm²

Draw a ring around the surface area of the pyramid.

61 cm²   100 cm²   124 cm²   136 cm²

[1]

14 Draw lines to match each equation to the sketch of its graph.

$x + y = 4$

$y = 2x - 3$

$y = \frac{x}{2} + 3$

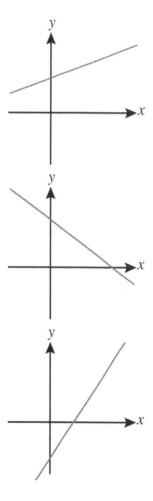

[1]

15 A cylinder has a radius of 4.5 cm and a height of 11.4 cm.

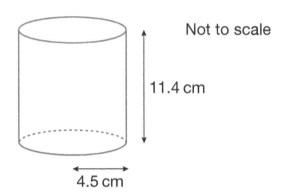

Not to scale

11.4 cm

4.5 cm

Find the volume of the cylinder.

_____ cm³

[2]

16 A straight line graph is shown on the grid.

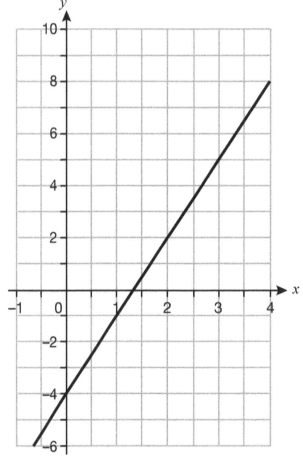

Find the equation of the line.

$y =$ _____

[2]

17 The pie chart shows information about the colours of flowers growing in a garden.

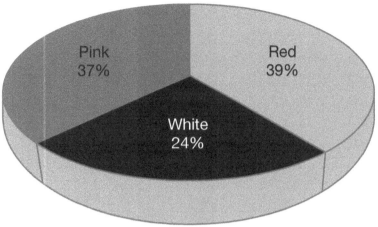

Explain why the pie chart is misleading.

_____

_____

[1]

18 Lucy has a fair spinner and an ordinary dice.
The spinner is numbered 1, 2, 3, 4 and 5
Lucy spins the spinner and throws the dice.

Here are four events.

K	L	M	N
Lucy gets the number 2 on the spinner.	Lucy gets an odd number on the spinner.	Lucy gets the number 2 on the dice.	Lucy gets an odd number on the dice.

Tick (✓) to show if the following pairs of events are independent or not independent.

Events	Independent	Not independent
K and M	☐	☐
M and N	☐	☐
L and M	☐	☐

[1]

19 The graph shows the relationship between the mass of cheese (in kg) and its cost (in $).

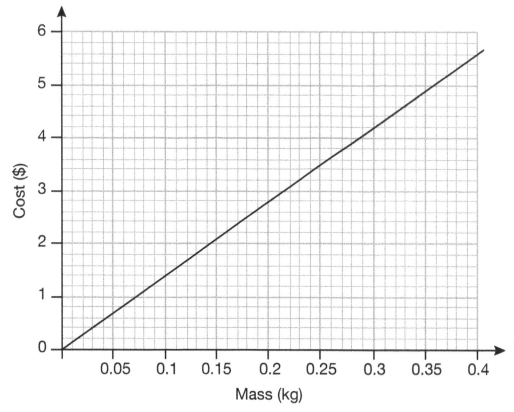

Find the price of 1 kg of this cheese.

$ _____

[2]

20 A 3D shape has exactly 4 planes of symmetry.

Describe or draw a possible shape that matches this description.

[1]

21 Max has four numbered tiles.

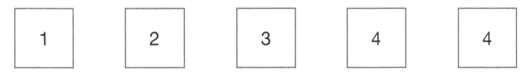

| 2 | 3 | 3 | 4 |

Sophia has five numbered tiles.

| 1 | 2 | 3 | 4 | 4 |

Max picks one of his tiles at random and Sophia picks one of her tiles at random.

Find the probability that both Max and Sophia pick a tile numbered with a 4

_____

[2]

22 Find the gradient of the line with equation $2x + 7y = 14$

_____

[2]

23 The diagram shows a solid triangular prism.

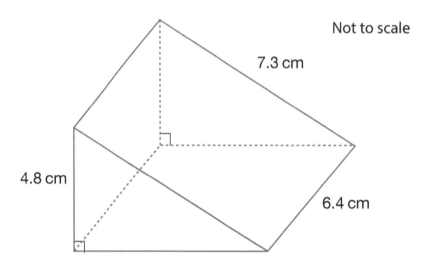

Not to scale

7.3 cm

4.8 cm

6.4 cm

Calculate the surface area of the prism.

_____ cm²

[3]

Total marks: $\dfrac{}{40}$

Assessment Task 6: Self-assessment

Enter the marks for each question in the unshaded cells below.

Question	Relationships and graphs	Probability 2	3D shapes	Simultaneous equations	Thinking statistically
1					
2					
3					
4					
5					
6					
7					
8					
9					
10					
11					
12					
13					
14					
15					
16					
17					
18					
19					
20					
21					
22					
23					
Total	/13	/9	/9	/6	/3

Some of the questions test your skills at Thinking and Working Mathematically. Write your marks for these questions in the grid below.

Question number	5	9	14	17	18	20	Total
Thinking and working mathematically							/7

The areas of the test that I am pleased with are

_____

_____

_____

The areas of the test that I found harder are

_____

_____

_____

Set yourself TWO targets.

TARGET 1

_____

TARGET 2

_____

_____

End of Book Test: Paper 1

Answer **all** questions.
Total marks for this paper: 50
You will need mathematical instruments for this test.
You may find tracing paper useful.

Calculators not allowed

1 Draw a ring around the calculation that has an answer greater than 15.

$$15 - \frac{1}{3} \qquad 15 \times \frac{1}{3} \qquad 15 \div 1\frac{1}{3} \qquad 15 \times 1\frac{1}{3}$$

[1]

2 Draw a ring around the set of values for which the inequality
$-2 < x + 4 < 6$ is true.

$$-2 < x < 2 \qquad -6 < x < 2 \qquad -6 < x < 6 \qquad 2 < x < 10$$

[1]

3 Draw a ring around the measurement equal to 0.3 millimetres.

300 micrometres   3000 micrometres   300 nanometres   3000 nanometres

[1]

4 Draw a ring around the value that is equal to $(-0.2) \times (-0.4)$

$$-0.8 \qquad -0.08 \qquad 0.08 \qquad 0.8$$

[1]

5 Simplify $n^3 \times n^3$

Draw a ring around your answer.

$$n^6 \qquad n^9 \qquad 2n^3 \qquad n^{27}$$

[1]

6 The back-to-back stem-and-leaf diagram shows the number of milkshakes that a café sold on 12 Mondays and on 12 Tuesdays.

	Monday				Tuesday			
9	5	3	**1**	5			
7 5 5	2	0	**2**	1	5	8	9
	6	1	**3**	0	2	4	6 6
	4	3	**4**	1	4		

Key: 3 | **1** | 5 represents 13 milkshakes sold on a Monday and 15 milkshakes sold on a Tuesday.

Draw a ring around the range of the number of milkshakes sold on these Mondays.

24   25   30   31

[1]

7 A fraction has a numerator of 7.

$$\frac{7}{\boxed{}}$$

Write a whole number greater than 7 in the denominator so that the fraction is equivalent to a recurring decimal.

[1]

8 Find the value of $\dfrac{37 - x^2}{2}$ when $x = 5$

[2]

9   The frequency polygon shows information about the ages of people living on Street 1.

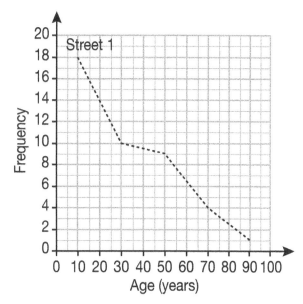

(a)   Information about the ages of people living on Street 2 is shown in the table.

Age (A years)	Frequency
$0 \leqslant A < 20$	6
$20 \leqslant A < 40$	14
$40 \leqslant A < 60$	10
$60 \leqslant A < 80$	8
$80 \leqslant A < 100$	4

On the grid above, draw a frequency polygon showing the ages of people living on Street 2.

[2]

(b)   Write one comparison of the ages of the people living on the two streets.

_____

_____

[1]

10 The diagram shows a triangle P drawn on a grid.

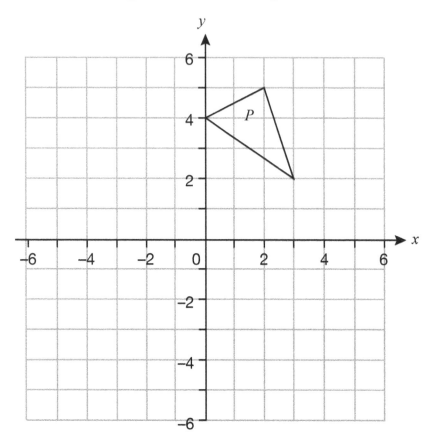

Triangle P is rotated by 180°, centre (0, 0), to give triangle Q.

Triangle Q is translated by vector $\begin{pmatrix} -1 \\ 5 \end{pmatrix}$ to give triangle R.

Draw triangle R on the grid.

[2]

11 Match each calculation to its answer.

60×10^2	6
600×10^{-2}	6000
$0.6 \div 10^{-3}$	600

[1]

12 (a) Complete the table of values for the function $2x + y = 8$

x	−1	0	1	2	3	4
y		8		4		

[2]

(b) Draw the graph of $2x + y = 8$ for values of x between −1 and 4

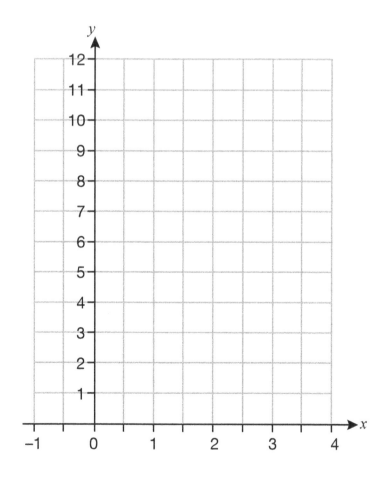

[2]

13 (a) Draw a ring around the two integers that are closest in value to $\sqrt{14}$

    2   3   4   5   6   7

[1]

(b) Allegra and Hector both estimate $\sqrt[3]{30}$

Allegra's estimate:
5.5

Hector's estimate:
3.1

Tick (✓) the name of the person who has the best estimate of $\sqrt[3]{30}$

Allegra ☐   Hector ☐

Give a reason for your answer.

_____

_____

[1]

14 AD and EH are parallel lines.
FJ and KC are parallel lines.

Angle $JBC = 60°$ and angle $FKG = 105°$

Find the value of x.

_____

[2]

15 Calculate $\frac{1}{2^3}$ × 17.6 × 80

_____

[2]

16 The term-to-term rule of a sequence is: 'square and then subtract 6'.

Here are three statements.

A	B	C
If the first term is 4, all terms in the sequence will be even numbers.	If the first term is 1, all terms in the sequence after the first term will be negative.	If the first term is 3, all terms in the sequence are equal.

Draw lines to show if each statement is true or false.

| Statement A |

| Statement B |

| Statement C |

| True |

| False |

[2]

17 Construct an angle of 45°
You may use the line below to help you.
You must show your construction arcs.

————————————————

[3]

18 Stella has a biased six-sided dice.
The table shows the probabilities that the dice lands on some of the numbers.

Number	1	2	3	4	5	6
Probability	0.20	0.22	0.10			

The three remaining probabilities in the table are equal.

Complete the table.

[2]

19 (a) Write 604 200 in standard form.

_____ [1]

(b) Write 7.2×10^{-3} as an ordinary number.

_____ [1]

20 The cross-section of a prism is a trapezium.

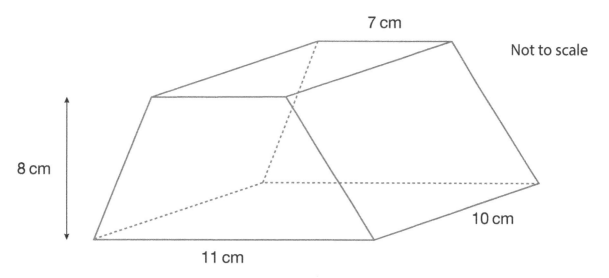

7 cm

Not to scale

8 cm

10 cm

11 cm

Find the volume of the prism.

_____ cm³

[2]

21 (a) Fill in the missing term in this expansion.

$(x + 5)(x - 11) = x^2$ _____ $- 55$

[1]

(b) Fill in the missing two terms in this expansion.

$(y - 6)^2 = y^2$ _____ _____

[2]

(c) George tries to expand $(z + 3)(z - 3)$

His expansion is:

$(z + 3)(z - 3) = z^2 - 6$

George has made a mistake.

Write the correct expansion of $(z + 3)(z - 3)$

_____

[1]

22 Priya has 8 cards.

She picks one of her cards at random and then replaces it.
She then picks a second card at random.

Find the probability that both of Priya's cards contain a quadrilateral.

_____

[2]

23 A rectangle is enlarged to make a larger rectangle.
Some information about the two rectangles is shown in the diagram.

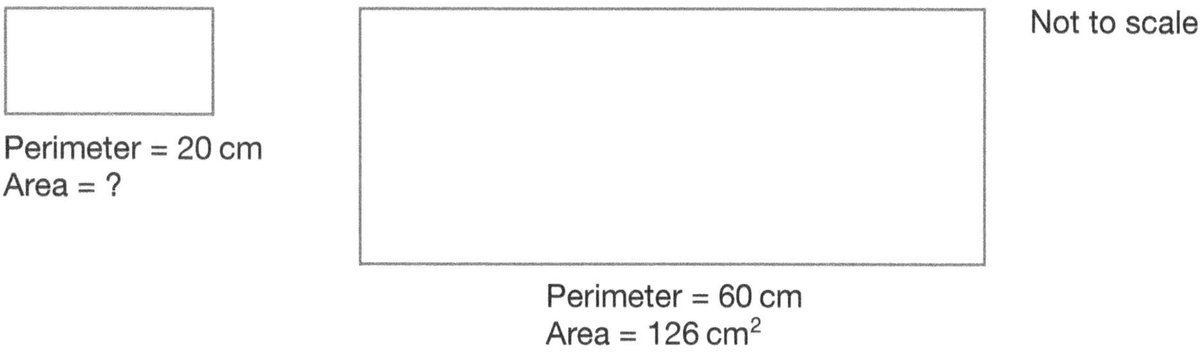

Not to scale

Perimeter = 20 cm
Area = ?

Perimeter = 60 cm
Area = 126 cm²

Find the area of the smaller rectangle.

_____ cm²

[2]

24 Find the value of $\left(2\frac{1}{4} - \frac{7}{12}\right) \div \left(\frac{4}{9} + \frac{2}{3}\right)$

Give your answer in its simplest form.

_____

[3]

Here are some simultaneous equations.

$3x + 4y = 8$

$4x + 5y = 12$

By solving the equations, show that $\frac{x}{y}$ is an integer.

[3]

Total marks: $\dfrac{}{50}$

End of Book Test: Paper 2

Answer **all** questions.
Total marks for this paper: 50
You will need mathematical instruments for this test.

Calculators allowed

1 The output, y, of a function when the input is x is given by the formula

$$y = 5x^2$$

Draw a ring around the output when the input is –2

    –100   –20   20   100

[1]

2 Draw a ring around the number that is not a rational number and is not an irrational number.

    $\sqrt{4}$ $\sqrt{\frac{1}{4}}$ $\sqrt{5}$ $\sqrt{-5}$

[1]

3 Draw a ring around the sum of the interior angles in a regular 9-sided polygon.

    180°   360°   1260°   1620°

[1]

4 The temperature of a hot drink is 60°C to the nearest 10°C.

Draw a ring around the lower bound for the temperature.

    50°C   55°C   59°C   59.5°C

[1]

5 Draw a ring around the solution to the equation $\frac{15}{w} = 3$

    $w = 5$ $w = \frac{1}{5}$ $w = 45$ $w = \frac{1}{45}$

[1]

6 The diagram shows two triangles drawn on a grid.

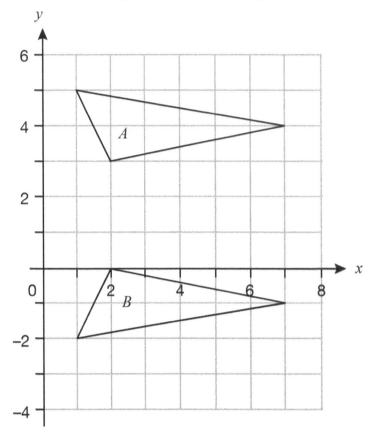

Draw a ring around the type of transformation that maps triangle A onto triangle B.

translation   enlargement   rotation   reflection

[1]

7 Find the value of m.

$$8^6 \times 8^{-2} = 8^m$$

$m =$ _____

[1]

8   The diagram shows some of the exterior angles of a pentagon.

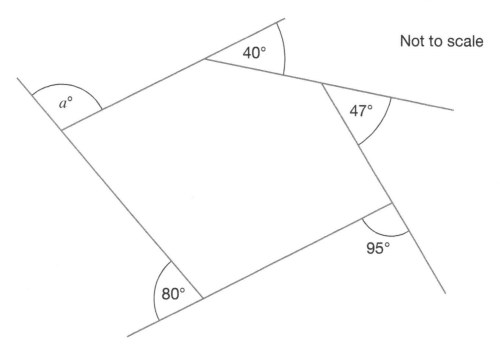

Not to scale

Calculate the value of a.

$a = $ _____

[2]

9 **(a)** Draw lines to match each description on the left to the expression on the right.

| I think of a number n. |
| I multiply by 5 |
| I then square. |
| Finally, I add 2 |

$(5n)^2 + 2$

| I think of a number n. |
| I multiply by 5 |
| I then add 2 |
| Finally, I square. |

$5(n + 2)^2$

| I think of a number n. |
| I add 2 |
| I then square. |
| Finally, I multiply by 5 |

$(5n + 2)^2$

[1]

(b) The base of a triangle has a length of $2x$ cm.

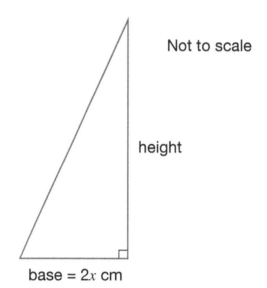

Not to scale

height

base = $2x$ cm

The height of the triangle is 3 times the base.

Find an expression for the area of the triangle.
Simplify your answer.

_____cm²

[2]

10 The table shows some information about the height of 10 horses.

Height, h (cm)	Frequency	Midpoint	Midpoint × frequency
$150 \leqslant h < 154$	1	152	152 × 1 = 152
$154 \leqslant h < 158$	2	156	156 × 2 = 312
$158 \leqslant h < 162$	4		
$162 \leqslant h < 166$	3		

By completing the table, find an estimate of the mean height of the horses.

_____cm

[3]

11 The diagram shows a cylinder with a radius of 7.8 cm and a height of 17.5 cm.

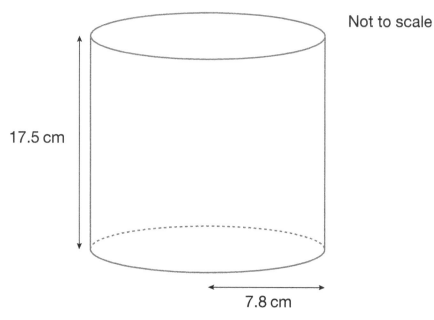

Not to scale

17.5 cm

7.8 cm

Calculate the volume of the cylinder.

_____cm³

[2]

12 (a) The nth term rules for two sequences are shown below.

Sequence 1
$n^2 - 2$

Sequence 2
$n^2 + 2$

Tick (✓) to show if each number is a term in Sequence 1 or a term in Sequence 2

	Sequence 1	**Sequence 2**
11	☐	☐
14	☐	☐
23	☐	☐
38	☐	☐

[2]

(b) A different sequence begins 1, 8, 27, 64, . . .

Write down a formula for the nth term of this sequence.

_____

[1]

13 A spinner has 4 sections, numbered 1, 2, 3 and 4
The table shows the probability that the spinner lands on each section when it is spun.

Section	1	2	3	4
Probability	0.15	0.2	0.25	0.4

The spinner is spun 300 times.

Estimate how many times the spinner is expected to land on either a 3 or a 4

_____

[2]

14 The diagram shows two right-angled triangles.

Not to scale

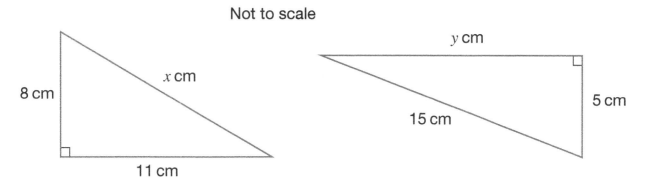

8 cm

x cm

11 cm

y cm

5 cm

15 cm

Show that $x < y$.

[3]

15 The grid shows a straight line.

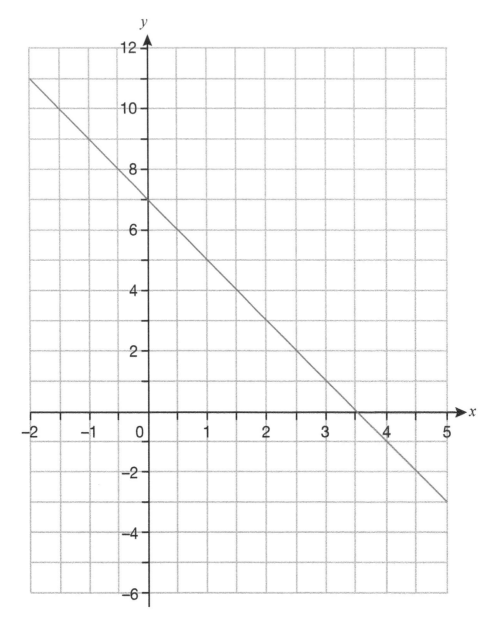

Find the equation of the straight line.

_____

[2]

16 The scale drawing shows the position of two towns, A and B.

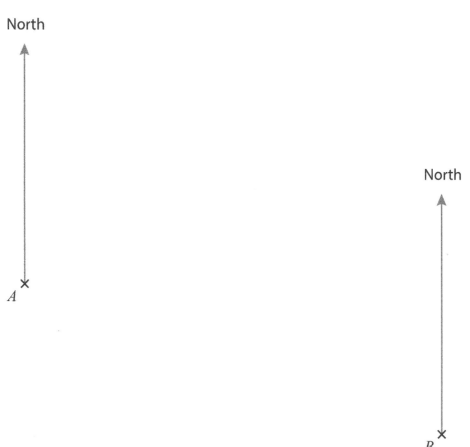

North

North

A

B

Scale: 1 cm represents 2 km

A third town C is
- on a bearing of 140° from A
- on a bearing of 270° from B.

Use the scale drawing to find the distance in real-life from town A to town C.
You should show the position of C on the drawing.

_____ km

[3]

17 Make g the subject of the formula $f = \frac{ag}{4} - 9$

_____

[2]

18 Paulo wants to find out what people living in a village think about the shops in the village centre.

Information about the population of the village is shown in the table.

Age group	Percentage of the population
Under 30 years old	40%
Over 30 years old	60%

(a) Paulo chooses a sample containing:
150 people aged under 30 years old
and
50 people aged over 30 years old.

Explain why Paulo's sample may give biased results.

_____

_____

[1]

(b) Complete the table to show how Paulo should obtain a sample of 200 people from the village.

Age group	Number of people in the sample
Under 30 years old	
Over 30 years old	
	Total = 200

[1]

19 The grid shows a kite K.

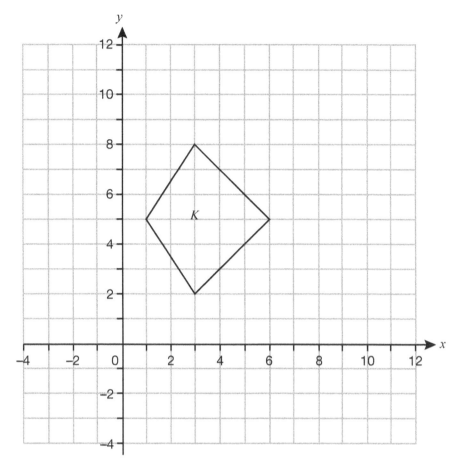

Enlarge kite K by scale factor 2, centre (4, 6)

[2]

20 Laila earns $18 000 per year.

Her pay is expected to increase by 3.5% each year.

Find how much Laila expects to earn per year after 4 years.

$_____

[2]

21 A tap fills up a container with water at a constant rate.
The container can hold 8.4 litres of water.
The graph shows how the volume of water in the container changes with time.

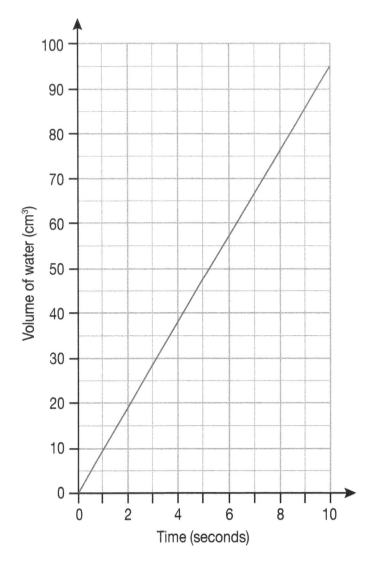

Calculate how long it takes for the tap to completely fill the container.
Give your answer in seconds.
(1 litre = 1000 cm³)

_____ seconds

[2]

22 The point A has coordinates (–17, 6) and B has coordinates (3, –2).

C is the point $\frac{1}{4}$ of the way along the line segment AB.

D is a point vertically above C.

Find the coordinates for one possible position of point D.

(_____ , _____)

[2]

23 Solve the inequality $9 - 2x \geqslant x + 15$

_____

[2]

24 Theo, Alex and Samira share a sum of money in the ratio 5 : 2 : 3
Samira receives $36

Theo gives 40% of his share to a charity.
Alex gives 25% of his share to the charity.
Samira gives all of her share to the charity.

Calculate how much money the charity receives in total.

$_____

[3]

25 The diagram shows a circle drawn inside a square.

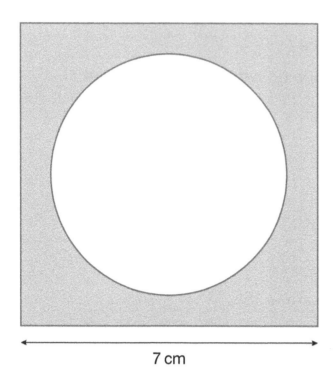

Not to scale

7 cm

Exactly half of the square is shaded.

Calculate the circumference of the circle.

_____ cm

[3]

Total marks: $\dfrac{}{50}$

End of Book Test: Self-assessment

Enter the mark for each question for Paper 1 and Paper 2 in the unshaded cells.

Paper 1

Question	Number	Algebra	Geometry and Measure	Statistics and Probability
1				
2				
3				
4				
5				
6				
7				
8				
9				
10				
11				
12				
13				
14				
15				
16				
17				
18				
19				
20				
21				
22				
23				
24				
25				
Total P1	/13	/17	/12	/8

Paper 2

Question	Number	Algebra	Geometry and Measure	Statistics and Probability
1				
2				
3				
4				
5				
6				
7				
8				
9				
10				
11				
12				
13				
14				
15				
16				
17				
18				
19				
20				
21				
22				
23				
24				
25				
Total P2	/8	/16	/19	/7

Overall total Paper 1 + Paper 2:

Total	/21	/33	/31	/15

Total mark: _____ /100

Thinking and working mathematically

Some of the questions test your skills at Thinking and Working Mathematically.
Write your marks for these questions in the grids below.

Paper 1

Question number	7	11	13(b)	16	21(c)	25	Total
Thinking and working mathematically							/9

Paper 2

Question number	9(a)	12(a)	14	18(a)	18(b)	22	Total
Thinking and working mathematically							/10

Overall total: _____ /19

The areas of the test that I am pleased with are

_____

_____

_____

The areas of the test that I found harder are

_____

_____

_____

Set yourself THREE targets.

TARGET 1

_____

TARGET 2

_____

TARGET 3

_____

Mark scheme Task 1

Question	Answer	Mark	Part marks
1	4.2×10^3	1	
2	5	1	
3	–5	1	
4	80×10^{-1} — 80; $0.8 \div 10^{-2}$ — 8; $8 \div 10^3$ — 0.008 (lines: 80×10^{-1} → 8, $0.8 \div 10^{-2}$ → 80, $8 \div 10^3$ → 0.008)	1	
5	9.2×10^4 (smallest) 7.5×10^5 8×10^5 (largest)	1	
6	(1) $\frac{1}{4}$ $\frac{1}{16}$	1	Both correct for 1 mark. Do not accept indices in the answer, for example $\frac{1}{4^1}$
7	Statements correctly completed using whole numbers. For example: $\sqrt{4}$ is a rational number $\sqrt{2}$ is an irrational number	1	Both statements should be correctly completed for 1 mark. Accept any correct answers, for example: Rational numbers: $\sqrt{9}$, $\sqrt{1}$, $\sqrt{144}$ ... Irrational numbers: $\sqrt{3}$, $\sqrt{11}$, $\sqrt{119}$ ...
8(a)	4.67×10^{-3}	1	
8(b)	5 065 000	1	
9(a)	22	1	
9(b)	(see table below)	2	Award 1 mark for 3, 4 or 5 values correctly added to the table.

Question 9(b) table:

	Cars		Not cars	
	Red	Not red	Red	Not red
Old	3	19	(1)	5
Not old	(4)	7	2	9

Question	Answer	Mark	Part marks
10	An answer suggesting Laxmi's answer is too big. For example • $\sqrt{45}$ is less than 7 • 7.1^2 is more than 49 (so her estimate is too big) • $\sqrt{45}$ is about 6.7 (accept 6.5 to 6.9)	1	Do not accept: • Laxmi's estimate is about right. • Laxmi's estimate is not very good.
11	750 000	2	Award 1 mark for $m = 750$ or $n = -3$ **or** for finding *their* $m \times 1000$
12	9.6×10^{-2}	1	
13	$\pi \times 2$	1	
14	115°	1	
15(a)	An answer that suggests that most of the table will not be used. For example • Three of the frequencies will be 0 • Only the first two intervals are needed • The highest value in the data is $35 • Nearly all the data is recorded in the first interval	1	
15(b)	A table which has equal class widths, no gaps and can be used to record all the data. At least three of the intervals should have non-zero frequencies. For example: 	Donation value ($)	Frequency
---	---
0–9.99	
10–19.99	
20–29.99	
30–39.99			1	Ignore any frequencies. Allow use of inequality notation.

Question	Answer	Mark	Part marks
16	157.5(°)	2	Award 1 mark for $\frac{360}{16}$ or 22.5(°) **or** for $(16 - 2) \times 180$ or 2520(°)
17		1	
18	($t =$) 128	3	Award 2 marks for forming a correct equation, for example • $125 + 100 + 111 + 3t = 720$ • $55 + 80 + 69 + 3u = 360$ (u = exterior angle to t) 2 marks can be implied by $u = 52$ **or** Award 1 mark for $(6 - 2) \times 180$ or 720 **or** for finding 80 and 69
19	A correct explanation. For example, • His customers are not likely to be representative of people in the town • He is only asking people who already use his shop • He should ask some people who are not his customers	1	Accept Not all of his customers will live in the town.
20	12(°)	3	Award 1 mark for sight of 60(°) **and** Award 1 mark for any of the following: • Angle AFE or angle GFC = 72(°) • Angle FCD or angle EFC or angle GFA = 108(°) • Angle FAE = 48(°)

Mark scheme Task 2

Question	Answer	Mark	Part marks
1	(−1, 1)	1	
2	111.5 grams	1	
3	8	1	
4	✓ ☐ / ☐ ✓ / ✓ ☐	1	
5	x^8	1	
6	15.822	2	Award 1 mark for digits 15822 **or** Award 1 mark for correct method to multiply with correct placement of decimal point in their answer so that it has 3 decimal places
7		2	Award 1 mark for U seen correctly on grid **or** Award 1 mark for correct rotation of *their* U (using correct centre and angle of rotation)
8	✗ | $x^2 + 13x + 42$ ✓ | ✗ | $z^2 - 2z - 35$	2	For 2 marks, ticks and crosses must be correct and expressions must be simplified. **or** Award 1 mark for an expression equivalent to $x^2 + 13x + 42$ or for an expression equivalent to $z^2 - 2z - 35$ (may be unsimplified).

Question	Answer	Mark	Part marks
9(a)	**Road A** Stem-and-leaf: 4 | 4 5 | 9 5 6 | 8 7 6 5 4 2 7 | 8 6 5 4 3 1 8 | 5 3 0 9 | 4 1	2	For 2 marks, the rows must be ordered. Award 1 mark for three completely correct ordered rows **or** if all values are in correct rows but rows are not ordered
9(b)	6.1 (m)	1	
9(c)	Houses are on average wider on Road A.	1FT	Follow through from their 9(b). Answers should be written in context, using vocabulary like 'houses' and 'wider', and should include an interpretation of median (for example, 'on average', 'generally').
10	$(y + 4)^2$	1	
11	5	1	
12	y^{10}	1	
13(a)	Reflection in the line $x = 1$	2	More than one transformation given scores 0 marks. Award 1 mark for 'reflection' **or** $x = 1$
13(b)	✓ ✓	1	
14		2	Ignore any lines to the left of 2.25 and to the right of 3.75 Award 1 mark for at least 3 points correctly plotted at midpoints (implied by lines). **or** Award 1 mark if *their* frequency polygon is shifted horizontally (for example if frequencies are consistently plotted at upper class boundaries).
15	$5 - 2x$	1	

Question	Answer	Mark	Part marks
16		2	Award 1 mark for 3 vertices correctly plotted **or** for image the correct size and orientation but incorrectly placed on grid.
17(a)	12.45 (cm)	1	
17(b)	7.95 (cm)	1	
18	Correct demonstration that area of M is more than 1000 (cm²). For example: $48 \times \left(\dfrac{12.5}{2.5}\right)^2 = 1200$ (cm²)	2	Award 1 mark for sight of: (scale factor =) $\dfrac{12.5}{2.5}$ or 5

Mark scheme Task 3

Question	Answer	Mark	Part marks
1	$\frac{3}{4}$	1	
2	$\frac{23}{30}$	1	
3	15	1	
4(a)	30	1	
4(b)	Ticks 'The dice is unlikely to be fair' **and** gives a correct reason. For example: • 10 is a lot less than 30 • The dice landed on 5 much less than expected • The relative frequency is $\frac{10}{180}$ and this is a lot less than $\frac{1}{6}$	1FT	Follow though from *their* 4(a) – accept the dice is likely to be fair if *their* 30 is close to 10 and a correct reason is given. Accept, for example, 10 is less than 30 Don't except reasons like • 10 is not 30 • The dice should have landed on 5 exactly 30 times
5	$\frac{5}{8}$	2	Award 1 mark for any of • answer of $\frac{15}{24}$ or equivalent unsimplified fraction • $\frac{3}{4} \times \frac{5}{6}$ • $\frac{15}{20} \times \frac{24}{20}$ or equivalent
6(a)	E located accurately on scale drawing (9.4 cm from D on a bearing of 096° from D) North North 96° D C E	2	Allow tolerance of ± 2 mm for length DE and allow tolerance of ± 2° for the bearing. Construction arc not necessary. Condone lack of label for E if intended position is clear. Award 1 mark if their E is 9.4 cm from D **or** if a line is drawn at a bearing of 096° from D.
6(b)	An answer in the range 138(°) to 142(°)	1FT	Follow through from *their* E.

Question	Answer	Mark	Part marks
7	$g = \dfrac{6h + 7}{5}$ or $g = \dfrac{6h}{5} + \dfrac{7}{5}$	2	Award 1 mark for answer $\dfrac{6h + 7}{5}$ or equivalent **or** correct first step in the rearrangement, for example • $6h = 5g - 7$ • $h + \dfrac{7}{6} = \dfrac{5g}{6}$
8	Complete method leading to answer 4	3	Award 2 marks for $\dfrac{16}{9} \times \dfrac{9}{4}$ or better **or** for correct calculation and simplification of (*their* $\dfrac{16}{9}$) $\times \dfrac{9}{4}$ provided their $\dfrac{16}{9}$ is from a valid method for subtracting fractions **or** Award 1 mark for sight of $1\dfrac{7}{9}$ or equivalent **or** converting two of $3\dfrac{2}{3}$, $1\dfrac{8}{9}$ and $2\dfrac{1}{4}$ to an improper fraction **or** $\dfrac{1}{9} + 1\dfrac{2}{3}$ or better **or** $2 + \dfrac{6}{9} - \dfrac{8}{9}$ or better **or** $1 + \dfrac{5}{3} - \dfrac{8}{9}$ or better
9	$\dfrac{2}{5}$	1	
10	(3, 7)	1	
11	$x = (y - 2)^2$	1	

Question	Answer	Mark	Part marks
12(a)	Correct completion of construction.	1	The construction arc should be seen.
12(b)	Correct completion of construction of an equilateral triangle. For example	2	Construction arcs must be seen for 2 marks. Award 1 mark for sight of at least two more arcs, radius 4 cm, centred on one of the intersection points
13	Recurring decimals: $\frac{8}{9}, \frac{4}{15}$ or $\frac{4}{9}, \frac{8}{15}$ **and** Terminating decimal $\frac{3}{12}$	2	Award 1 mark for making one proper fraction equivalent to a recurring decimal and one proper fraction equivalent to a terminating decimal with no repetition of numbers in these two fractions **or** Award 1 mark for forming 2 recurring decimals and 1 terminating decimal with repetition of numbers or with some improper fractions.
14(a)	Table with P(odd number) = 0.6 and sum of probabilities = 1. For example 	Number	1	2	3	4	 |---|---|---|---|---| 	Probability	0.5	0.2	0.1	0.2		1	

Question	Answer				Mark	Part marks
14(b)	**Colour**	Red	Green	Blue	2	Accept equivalent fractions or percentages.
	Probability	(0.28)	0.36	0.36		Award 1 mark for $\frac{1-0.28}{2}$ **or** for forming an equation, such as $x + x + 0.28 = 1$
15	−3				2	Award 1 mark for answer 3 or answer ±3 **or** for $(-)\sqrt{\frac{90}{10}}$ or better.
16	(8, 5)				2	Award 1 mark for one correct coordinate in answer **or** finding B as (12, 5) **or** finding C as (−4, 5)

Mark scheme Task 4

Question	Answer	Mark	Part marks
1	Class 4	1	
2	−4	1	
3	0, 1, 2	1	
4(a)	$4 \leqslant x < 6$	2	Award 1 mark for $\frac{40}{2}$ or 20, or $\frac{40+1}{2}$ or 20.5
4(b)	10 (hours)	1	
5	Correct demonstration that the triangle is not right-angled. For example: • $11^2 = 121$ **and** $5^2 + 9^2 = 106$ • $\sqrt{5^2 + 9^2} = \sqrt{106}$ which is not 11	2	Award 1 mark for sight of any two of 5^2, 9^2, 11^2 or equivalent. 1 mark is implied by sight of 106
6		1	
7(a)	A correct demonstration that $\frac{1}{5} + 1.8 \times 10$ is 18.2 or equivalent. For example: • She should have multiplied first. $\frac{1}{5} + 18 = 18\frac{1}{5}$ • $\frac{1}{5} + 1.8 \times 10 = 0.2 + 18 = 18.2$	1	Some working needs to be seen as well as the correct answer.
7(b)	0.72 or $\frac{18}{25}$	2	Award 1 mark for $4^2 \times \frac{1}{8} = 2$ (seen or implied) **or** $4^2 \times 0.36 = 5.76$ **or** $0.36 \times \frac{1}{8} = 0.045$ or equivalent **or** formulating as fractions with some cancelling, for example $\frac{36}{100} = \frac{9}{25}$ or $^816 \times \frac{36}{\cancel{100}_{50}} \times \frac{1}{8}$

Question	Answer	Mark	Part marks
8	($m =$) 6.5 or equivalent	3	Award 1 mark for $27 = 3(2m - 4)$ or better, for example $2m - 4 = \dfrac{27}{3}$ **and** Award 1 mark for correct 1st step in solving an equation of the form $am - b = c$ (adding b to both sides or dividing both sides by a).
9	$16 \leqslant x < 18$	1	
10	$r^2 = p^2 + q^2$	1	
11	$n = 4$	1	
12		1	All matched correctly.
13	14.4(2…)	2	Award 1 mark for 8^2 and 12^2 1 mark can be implied by sight of 208
14(a)	13.5 (m)	3	Award 2 marks for sight of 540 **or** Award 1 mark for multiplying at least one midpoint by its corresponding frequency **and** Award 1 mark for adding four terms of the form $x \times f$, where x is a number in the class interval (could be one of the class boundaries) and f is the corresponding frequency.
14(b)	A comparison on the mean values in context. For example, • (The mean for Park A is less than the mean for Park B meaning that) trees in Park A are shorter on average (than those in Park B). • The trees in Park B are generally taller.	1	The comparison should refer to trees as well as language such as taller/ shorter/ smaller/ larger.

Question	Answer	Mark	Part marks
15	$x < -2$	2	Award 1 mark for answer involving –2 (such as $x = -2$, $x > -2$ etc) **or** Award 1 mark for $11 - 23 > 2x + 4x$ or $-2x - 4x > 23 - 11$ or better (allow any inequality sign or equals).
16	66.1(2…) (cm)	3	Allow 3 marks for answer of 66 (cm) if working is seen. Award 1 mark for $(BC^2 =)$ $40^2 - 26^2$ or 924 or for $(BC =)$ $\sqrt{40^2 - 26^2}$ or 30.3(97…) or 30.4 **and** Award 1 mark for $(AD^2) = 26^2 + (2 \times their\ BC)^2$ or 4372 **or** $(AD) = \sqrt{26^2 + (2 \times their\ BC)^2}$

Mark scheme Task 5

Question	Answer	Mark	Part marks
1	500×1.04^3	1	
2	26	1	
3	$\pi \times 4^2$	1	
4	An accurate compound bar chart with • bars of equal width • bars not touching • accurate subdivision of bars • shading to match key. 	3	Cars and vans can be stacked either way round. Award 2 marks for one fully correct bar including shading **or** for two bars of equal width and accurate subdivision but may not be shaded (condone bars touching for 2 marks) **or** Award 1 mark for one bar of correct height and correct subdivision but with no or wrong shading **or** for two bars of equal width and correct total height with no or wrong subdivision (condone bars touching).
5	42 (seconds)	1	
6	<table><tr><th>Statement is correct</th><th>Statement is not correct</th></tr><tr><td>(A) C D</td><td> B E</td></tr></table>	2	Award 1 mark for 2 or 3 statements correctly added to table.
7(a)	A correct description of the relationship. For example: • As marks on the sports quiz increase, the marks on the history quiz also increase. • Students who score highly on the sports quiz tend to score highly on the history quiz. • The marks on the two quizzes are positively correlated.	1	The description should include some context. Just 'positive correlation' is insufficient for 1 mark.

Question	Answer	Mark	Part marks
7(b)	A suitable ruled line of best fit. For example:	1	The line of best fit should extend from at least 4.5 to 19.5 on horizontal axis. *Their* line should have a positive gradient. There should be at least 3 points on either side of *their* line of best fit.
7(c)	Correct reading from *their* line of best fit.	1FT	Allow answer to be a decimal or rounded to nearest whole number.
8	($) 864	3	Award 2 marks for correct method to find 0.9×960, e.g. <ul><li>$960 - 96$</li><li>1.08×800</li></ul> **or** Award 1 mark for any of the following <ul><li>sight of 960</li><li>correctly finding 90% of *their* 960</li><li>sight of 1.2×0.9 or 1.08 or 108% or 8% increase</li></ul>
9	500 (ml) **and** 180 (ml)	2	Award 1 mark for each correct value.
10	n^3 and $n^2 - 5$	3	Award 1 mark for n^3 or equivalent **and** either award 2 marks for $n^2 - 5$ or equivalent **or** award 1 mark for $n^2 (\pm ...)$ or equivalent
11	$980	1	
12	7	1	
13	Strong negative	1	
14	An answer between 94.9 and 95.1	1	Allow $\frac{121}{4}\pi$

Question	Answer	Mark	Part marks
15(a)	7000 (nanometres)	1	
15(b)	1 megabyte or 1 MB	1	
16	120 (hours)	1	
17	A correct pie chart, for example: 	3	Award 1 mark for one sector of 180° **and** Award 1 mark for a sector measuring either 99° or 81° (tolerance ±2°) **and** Award 1 mark for labelling *their* largest sector 'Art', *their* smallest sector 'Drama' and the remaining sector 'Music'.
18	An answer between 107.9 (cm) and 108 (cm)	2	Allow 3.14(2…) and $\frac{22}{7}$ for π Award 1 mark for $\frac{\pi \times 42}{2}$ or 65.9(4) to 66
19(a)	25 **and** 32	1	
19(b)	5	2	Award 1 mark for choosing a value for the 1st term and then using the term-to-term rule to find the 2nd and 3rd terms.
20	Correct working leading to an answer between 132 (cm²) and 132.1 (cm²). For example: $8.6^2 + \frac{13.2 \times 13.2}{2} - \pi \times \left(\frac{8.6}{2}\right)^2 \div 2$	3	Allow 3.14(2…) and $\frac{22}{7}$ for π Award 2 marks for $\pi \times \left(\frac{8.6}{2}\right)^2 \div 2$ or 29(.0…) **and** 8.6^2 **or** $\frac{13.2 \times 13.2}{2}$ **or** Award 1 mark for any of • $\pi \times \left(\frac{8.6}{2}\right)^2 \div 2$ or $\pi \times \left(\frac{8.6}{2}\right)^2$ or 29(.0…) or 58.0… to 58.1… • 8.6^2 or 73.96 • $\frac{13.2 \times 13.2}{2}$ or 87.12

Question	Answer	Mark	Part marks
21	An answer between ($)517.90 to ($)518	2	Award 1 mark for $950 \times 0.8 \times (1 - 0.12)^n$ where $n = 1, 2, 3$ or 4 **or** for correctly decreasing 950 by 20% (to get 760) and then correctly reducing an amount by 12% at least once.

Mark scheme Task 6

Question	Answer	Mark	Part marks
1	2	1	
2	3	1	
3	$\frac{1}{12}$	1	
4	180 (cm³)	1	
5	Two reasons from ● No scale on the vertical axis ● Horizontal scale is not linear ● Line width is too wide.	2	1 mark for each correct reason.
6	$(x =)$ 13 **and** $(y =)$ 2	2	Award 1 mark for $(x =)$ 13 **or** $(y =)$ 2 **or** Award 1 mark for $3x = 39$ **or** Award 1 mark for $3y = 6$
7(a)		2	Award 1 mark for 2 or 3 probabilities filled in correctly.
7(b)	$\frac{4}{35}$	2	Award 1 mark for $\frac{1}{5} \times \frac{4}{7}$
8(a)		2	Award 1 mark for 2 or 3 correct values.

For 7(a) tree diagram:
- pizza: $\frac{1}{5}$; reads $\frac{4}{7}$, does not read $\frac{3}{7}$
- not pizza: $\frac{4}{5}$; reads $\frac{4}{7}$, does not read $\frac{3}{7}$

For 8(a) table:

x	−2	−1	0	1	2	3	4
y	−2	(−5)	(−6)	−5	−2	(3)	10

Question	Answer	Mark	Part marks
8(b)	The graph of $y = x^2 - 6$ drawn accurately – points joined with a smooth curve.	2	Award 1 mark for plotting at least 5 of *their* points accurately – implied by a curve or line passing through them.
9	Ticks 'He is not correct' and gives a suitable reason explaining how the probability depends on the first ball picked. For example, • If the first ball is red, 2 (out of 4) balls left are red. But if the first ball is green, 3 (out of 4) balls are red. • The probability that second ball is red is less if the first ball was red than if the first ball was green. • If the first ball picked was red, there will be fewer red balls left in the bag.	1	Saying 'the probability depends on the colour of the first ball picked' is not quite sufficient.
10	132 (cm²)	3	Award 2 marks for either $y = -2$ **or** for $x = 9$ **or** for 12×11 **or** Award 1 mark for forming the equations $2x + 3y = 12$ and $x + 2 = 9 - y$ **or** for eliminating one of the variables from their equations.

Question	Answer	Mark	Part marks
11	$x = 3$, $y = 4$	1	
12	$4x + 10y = 72$	1	
13	$136\,\text{cm}^2$	1	
14		1	
15	An answer in the range 724 to 726 (cm³)	2	Award 1 mark for $\pi \times 4.5^2$ ($\times 11.4$) Allow 3.14(2…) or $\frac{22}{7}$ for π 1 mark can be implied by 63.5… to 64
16	$y = 3x - 4$ or equivalent	2	Award 1 mark for gradient = 3 (implied by $3x$ in answer) **or** for answer of the form $y = mx - 4$ ($m \neq 0$)
17	A suitable reason. For example, ● The 3D nature of the pie chart makes the sector for 'white' look larger than it should be. ● It is not easy to compare the sizes of the sectors.	1	Allow 'the graph is 3 dimensional.'
18	✓ ☐ ☐ ✓ ✓ ☐	1	

Question	Answer	Mark	Part marks
19	($)14 (accept answers between 13.80 and 14.20 inclusive)	2	Award 1 mark for evidence of a suitable strategy. For example, • Reading off at 0.25 and then multiplying the cost by 4 • Finding 2 × (reading at 0.4) + (reading at 0.2) • Reading off at 0.2 and then multiplying the cost by 5 • Reading off at 0.4 and then dividing by 0.4
20	Any indication of a square-based pyramid **or** a prism with an equilateral triangle as cross-section	1	Accept other possible answers, such as a prism with the following hexagonal cross-section
21	$\frac{2}{20}$ or $\frac{1}{10}$	2	Accept equivalents, such as 0.1 Award 1 mark for $\frac{1}{4} \times \frac{2}{5}$
22	$-\frac{2}{7}$	2	Award 1 mark for 1st step to make y the subject, for example $7y = 14 - 2x$ **or** for attempt to rearrange to the form $y = mx + c$ and gives gradient as their numerical m.
23	139(.04) (cm²)	3	Award 1 mark for $\sqrt{7.3^2 - 4.8^2}$ or 5.5 **and** Award 1 mark for correct working for the surface area 2 × (0.5 × 4.8 × *their* 5.5) + (7.3 × 6.4) + (6.4 × *their* 5.5) + (4.8 × 6.4)

Mark scheme Paper 1

Question	Answer	Mark	Part marks
1	$15 \times 1\frac{1}{3}$	1	
2	$-6 < x < 2$	1	
3	300 micrometres	1	
4	0.08	1	
5	n^6	1	
6	31	1	
7	Whole number greater than 7 written in denominator to give a fraction equivalent to a recurring decimal. For example: $\frac{(7)}{9}, \frac{(7)}{11}$, etc	1	
8	6	2	Award 1 mark for 25 or 12 seen.
9(a)	A correct frequency polygon for Street 2 	2	Condone lack of label for Street 2 if intended answer is clear. Award 1 mark for at least 3 of the following points plotted: (10, 6), (30, 14), (50, 10), (70, 8), (90, 4) (can be implied by line passing through these points) **or** Award 1 mark if their frequency polygon is a translation of the correct one (i.e. vertical heights correct but horizontal coordinates consistently incorrect).
9(b)	Any correct comparison of the ages of people on the two streets. For example, • More younger people living on Street 1 • Street 2 has a higher proportion of older people • Street 2 has more people aged $20 \leqslant A < 40$ • People are older on average on Street 2	1	

Question	Answer	Mark	Part marks
10		2	Award 1 mark if Q marked in correct position on grid **or** Award 1 mark if their Q is translated correctly by $\begin{pmatrix} -1 \\ 5 \end{pmatrix}$
11		1	Accept intention.
12(a)	<table><tr><td>x</td><td>−1</td><td>0</td><td>1</td><td>2</td><td>3</td><td>4</td></tr><tr><td>y</td><td>10</td><td>(8)</td><td>6</td><td>(4)</td><td>2</td><td>0</td></tr></table>	2	Award 1 mark for 2 or 3 values correct.
12(b)	A line joining (−1, 10) to (4, 0) 	2	Award 1 mark for plotting 4 of the points from *their* table (can be implied by a line passing through those points).
13(a)	3 **and** 4	1	

Question	Answer	Mark	Part marks
13(b)	Ticks Hector **and** gives a correct reason. For example: • $\sqrt[3]{27} = 3$ so $\sqrt[3]{30}$ must be a little bigger than 3 • $\sqrt[3]{30}$ must be less than 4 as $4^3 = 64$	1	
14	15	2	Award 1 mark for sight of any of these angles (check diagram): • Angle $BFG = 60°$ • Angle $BFK = 75°$ • Angle $BCG = 60°$ • Angle $FGK = 60°$
15	176	2	Award 1 mark for $\frac{1}{2^3} \times 80 = 10$ **or** for $\frac{1}{2^3} \times 17.6 = 2.2$ **or** for $17.6 \times 80 = 1408$ **or** for cancelling 2's in the denominator with numbers in the numerator, e.g. $\frac{1}{\cancel{2} \times \cancel{2} \times \cancel{2}} \times {}^{8.8}\cancel{17.6} \times {}^{20}\cancel{80}$
16	Statement A — True Statement B Statement C — False	2	Award 1 mark for 2 correct lines.

Question	Answer	Mark	Part marks
17	Accurate construction of an angle of 45° with arc lines seen, by the method shown below, or any other suitable construction. 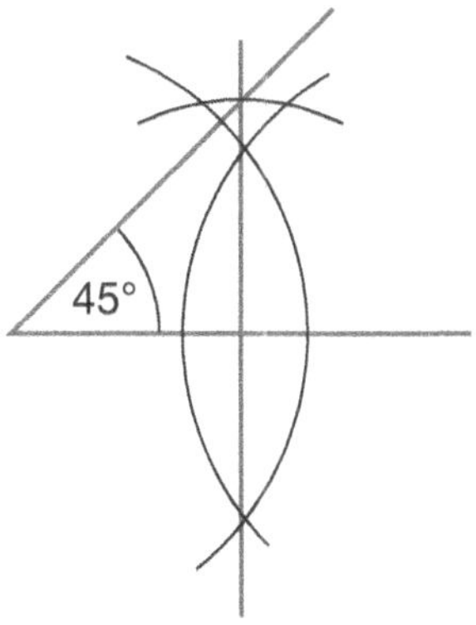 	3	Award 1 mark for construction of perpendicular bisector of original line (or other construction of a perpendicular line) **and** Award 1 mark for drawing an arc of radius 4 cm, centred on the midpoint of the base line, that intersects the perpendicular line **or** Award 1 mark for bisecting their 90° angle.
18	<table><tr><td>1</td><td>2</td><td>3</td><td>4</td><td>5</td><td>6</td></tr><tr><td>(0.20)</td><td>(0.22)</td><td>(0.10)</td><td>0.16</td><td>0.16</td><td>0.16</td></tr></table>	2	Award 1 mark for $1 - 0.2 - 0.22 - 0.1$ or 0.48
19(a)	6.042×10^5	1	
19(b)	0.0072	1	
20	$720\,\text{cm}^3$	2	Award 1 mark for $0.5 \times (7 + 11) \times 8$ or 72 **or** for *their* cross-sectional area $\times\ 10$
21(a)	$-6x$	1	
21(b)	$-12y + 36$	2	Award 1 mark for $-12y$ **or** for 36
21(c)	$z^2 - 9$	1	
22	$\dfrac{9}{64}$	2	Award 1 mark for $\dfrac{3}{8} \times \dfrac{3}{8}$ or equivalent.
23	$14\ (\text{cm}^2)$	2	Award 1 mark for $\left(\dfrac{60}{20}\right)^2$ or 9 or $\left(\dfrac{20}{60}\right)^2$ or $\dfrac{1}{9}$

Question	Answer	Mark	Part marks
24	$\frac{3}{2}$ or $1\frac{1}{2}$	3	Award 2 marks for an unsimplified fraction equivalent to $1\frac{1}{2}$ or $\frac{3}{2}$ **or** Award 2 marks for $\frac{5}{3}$ or equivalent improper fraction **and** $\frac{10}{9}$ or equivalent improper fraction or $\frac{9}{10}$ or equivalent **or** Award 1 mark for $\frac{5}{3}$ or $1\frac{2}{3}$ or equivalent fraction **or** $\frac{10}{9}$ or $1\frac{1}{9}$ or $\frac{9}{10}$ or equivalent fractions.
25	Correct algebraic working leading to $x = 8$ **and** $y = -4$ **and** $\frac{x}{y} = -2$	3	Award 2 marks for $x = 8$ **or** $y = -4$ **or** Award 1 mark for correct method to solve the simultaneous equations. For example, making the coefficients of x or y equal and then correct decision to add/ subtract to eliminate the variable. **or** Award 1 mark for a pair of values of x and y which satisfy one of the original equations.

Mark scheme Paper 2

Question	Answer	Mark	Part marks
1	20	1	
2	$\sqrt{-5}$	1	
3	1260°	1	
4	55°C	1	
5	$w = 5$	1	
6	reflection	1	
7	$(m =) 4$	1	
8	$(a =) 98$	2	Award 1 mark for: 360 − 40 − 47 − 95 − 80, or for $a + 40 + 47 + 95 + 80 = 360$ or equivalent.
9(a)		1	
9(b)	$6x^2$	2	Award 1 mark for $\frac{2x \times 6x}{2}$ or better.
10	159.6 (cm)	3	Award 1 mark for correct completion of remainder of table. | (152)	(152 × 1 = 152)
(156)	(156 × 2 = 312)
160	640
164	492	 **and** Award 1 mark for: $\dfrac{152 + 312 + \textit{their } 640 + \textit{their } 492}{10}$
11	An answer in the range 3340 to 3350 (cm³)	2	Allow 3.14(2) and $\frac{22}{7}$ for π. Award 1 mark for π × 7.8² (× 17.5) or for 191.(…)

Question	Answer	Mark	Part marks
12(a)	☐ ✓ ✓ ☐ ✓ ☐ ☐ ✓	2	Award 1 mark for 3 correct ticks.
12(b)	n^3 or equivalent	1	
13	195	2	Award 1 mark for 0.25 × 300 or 75 **or** 0.4 × 300 or 120 **or** (0.25 + 0.4) × 300
14	Correct working that demonstrates $x < y$. This could include • sight of $(x =) \sqrt{185}$ or 13.6(...) **and** $(y =) \sqrt{200}$ or 14.1(...) • sight of $x^2 = 185$ **and** $y^2 = 200$	3	Award 2 marks for $(x =) \sqrt{185}$ or 13.6(...) or $(y =) \sqrt{200}$ or 14.1(...) or $x^2 = 185$ or $y^2 = 200$ **or** Award 2 marks for sight of 185 and 200 **or** Award 1 mark for $8^2 + 11^2$ or $15^2 - 5^2$ or better.
15	$y = 7 - 2x$ or equivalent simplified equation.	2	Award 1 mark for correct method to find gradient, for example $\frac{-7}{3.5}$ or better (implied by answer of the form $y = -2x + c$) **or** Award 1 mark for answer of the form $y = mx + 7$ (for a numerical value of m)
16	Scale drawing showing point C accurately plotted and *their* diagram used to find the distance from A to C 	3	Allow tolerance of ±2° for each bearing and tolerance of ±2 mm for the measurement of AC. Expect an answer of about 10.6 (km) Award 2 marks for accurately finding the position of C. **or** Award 1 mark for accurately drawing either bearing **or** for accurate use of the scale to find the real distance AC from *their* C.

Question	Answer	Mark	Part marks
17	$g = \dfrac{4(f + 9)}{a}$ or equivalent	2	Award 1 mark for a correct first step in the rearrangement: • adding 9 to both sides • multiplying both sides by 4 • dividing both sides by a
18(a)	A suitable explanation. For example • Most people in the village are over 30 but in the sample most people are under 30 (and age may affect views about the shops) • He has asked too many people aged under 30 • He hasn't asked enough people aged over 30	1	Accept: 'His sample is not representative of the people living in the village.' Do not accept 'He has not asked enough people'.
18(b)	(Under 30 years old) 80 (Over 30 years old) 120	1	Both numbers correct in table.
19		2	Award 1 mark if 2 or 3 vertices of the enlarged kite are found correctly **or** for an image of the correct size and orientation but positioned incorrectly on the grid.
20	($)20 655(.41)	2	Award 1 mark for 1.035^4 or equivalent. **or** Award 1 mark for finding Laila's earnings after 1, 2 and 3 years (implied by sight of 18630, 19282.05 and 19956.92…)
21	An answer between 880 and 890 (seconds).	2	Award 1 mark for calculating gradient (for example $\dfrac{95}{10}$ or 9.5) **or** for finding 8400 ÷ *their* 9.5 **or** for finding $\dfrac{8400}{95}$ (× 10)

Question	Answer	Mark	Part marks
22	An answer of the form $(-12, y)$ where $y > 4$ For example, $(-12, 5)$	2	Award 1 mark for -12 **or** for $\frac{3-(-17)}{4}$ or 5 **or** for $\frac{-17-3}{4}$ or -5 **or** for finding y coordinate of C as 4
23	$x \leqslant -2$ or $-2 \geqslant x$	2	Award 1 mark for answer -2, perhaps with wrong inequality sign (for example, $x = -2$ or $x \geqslant -2$) **or** Award 1 mark for correct first step in solving the inequality (allowing any inequality sign or =), for example: • $9 \geqslant 3x + 15$ • $9 - 3x \geqslant 15$ • $-2x \geqslant x + 6$ • $-6 - 2x \geqslant x$
24	($)66	3	Award 2 marks for $0.4 \times \frac{36}{3} \times 5$ **and** $0.25 \times \frac{36}{3} \times 2$ or equivalent (implied by $\frac{36}{3} \times 5.5$ or $\frac{36}{3} \times 2.5$) **or** 24 and 6 **or** 30 **or** Award 1 mark for (Theo =) $\frac{36}{3} \times 5$ or 60 (implied by 24) **or** (Alex =) $\frac{36}{3} \times 2$ or 24 (implied by 6) **or** $0.4 \times 5 + 0.25 \times 2$ or 2.5 or 5.5 **or** $0.4 \times$ *their* $60 + 0.25 \times$ *their* $24 + 36$
25	An answer between 17.5 and 17.6 (cm)	3	Allow 3.14(2) and $\frac{22}{7}$ for π. Award 1 mark for $r = \sqrt{\dfrac{7^2 \div 2}{\pi}}$ or 2.79(...) or 2.8 **and** Award 1 mark for $2 \times \pi \times$ *their* 2.79(...)